A Watched Pot

How We Experience Time

MICHAEL G. FLAHERTY

NEW YORK UNIVERSITY PRESS
New York and London

NEW YORK UNIVERSITY PRESS
New York and London

Library of Congress Cataloging-in-Publication Data
Flaherty, Michael G.
A watched pot : how we experience time / Michael G. Flaherty.
p.   cm.
Includes bibliographical references and index.
ISBN 0-8147-2687-9 (alk. paper)
1. Time perception. 2. Time perspective. I. Title.
BF468 .F57 1998
153.7'53—ddc21   98-25523
                                CIP

New York University Press books are printed on acid-free paper,
and their binding materials are chosen for strength and durability.

Manufactured in the United States of America

10 9 8 7 6 5 4 3 2 1

For Robert and Constance Ann Flaherty

Contents

Acknowledgments

Increasingly, it seems, time is a precious resource. Nonetheless, throughout the course of this project, a great many people have given abundantly of their time by helping me in one way or another. Their generous efforts have made this project possible.

Much of this study has been shaped by the challenging but always benevolent advice of my friend and mentor, Norman Denzin. Likewise, I am grateful to Gary Alan Fine and Eviatar Zerubavel for their inspiration, insight, and counsel. I want to thank Gretchen Flaherty (she who alters my experience of time) as well as Lori Ducharme and Laurel Richardson for invaluable comments on earlier versions of this manuscript. In particular, I want to acknowledge Laurel's help with the title. For their enthusiasm and guidance, I am indebted to Karen Cook, Carolyn Ellis, Keith Irwin, James MacDougall, and David Maines. When the chips were down, Joel Best, Peter Hall, Jack Katz, Virginia Olesen, and Clint Sanders were willing to serve as reviewers. Special thanks go to Jack Katz and Barry Glassner for giving this project the boost it needed at a critical juncture.

Linda O'Bryant typed every word of this manuscript, and, with a wicked sense of humor, indulged me in my efforts to get it right. Leslie Wasson, Diane Mennella Hood, and Michele Buheit helped with the interviewing and coding, while Michelle (Mikki) Meer worked with me on the research for chapter 5. I also appreciate the support and assistance of Howard Carter, Lloyd Chapin, Edmund Gallizzi, Jamie Hastreiter, David Henderson, Catherine McCoy, Melissa Merrick, Lisa Mills, Kate Nangeroni, Janice Pelzer, and Elizabeth (Betsy) Wilson. Finally, for initial interest, repeated consultations, and patience, I want to thank Tim Bartlett, my editor, and Michael Rose, his editorial assistant.

Portions of this book have been previously published in slightly different form: "The Neglected Dimension of Temporality in Social Psychology," in *Studies in Symbolic Interaction* 8 (1987): 143–55, with permission of JAI Press; "Multiple Realities and the Experience of Duration," in *The Sociological Quarterly* 28 (1987): 313–26, with permission of the Midwest Sociological Society; "The Perception of Time and Situated Engrossment," in *Social Psychology Quarterly* 54 (1991): 76–85, with permission of the American Sociological Association; "The Erotics and Hermeneutics of Temporality," in *Investigating Subjectivity*, edited by Carolyn Ellis and Michael G. Flaherty, 141–55 (Newbury Park, Calif.: Sage, 1992), with permission of Sage Publications; "Conceptualizing Variation in the Experience of Time," in *Sociological Inquiry* 63 (1993): 394–405, with permission of the University of Texas Press; "Some Methodological Principles from Research in Practice: Validity, Truth, and Method in the Study of Lived Time," in *Qualitative Inquiry* 2 (1996): 285–99, with permission of Sage Publications; with Michelle D. Meer, "How Time Flies: Age, Memory, and Temporal Compression," in *The Sociological Quarterly* 35 (1994): 705–21, with permission of the Midwest Sociological Society.

A Watched Pot

[1]

Introduction

AS YOU READ these words, consider the following fact: that in so doing, you are marking time. There is no need to consult your watch; its movements merely externalize and run parallel to the kaleidoscopic stream of experience that you distill into an image of duration. You inhale and exhale as the blood courses through your veins. You shift the position of your legs, and, hearing something, look up in time to see a petal drop from the flowers on your table. In like fashion, these words enter your stream of consciousness as a skein of events.[1]

The fundamental quality of duration is embedded in our impression that things are not as they were before; that is, that things have changed.[2] Human beings are unique in part because they are capable of fusing their experience of heterogeneous events into a coherent sense of persistence. We can attend to change by remembering the past, stepping back from the present, and anticipating the future. Our clocks and calendars mark time, but they do not make time.

Only human beings make time by sifting the fragmentary dynamics of experience through the reflexive "unity of consciousness."[3]

On its most basic level, then, temporality is an aspect of subjectivity. This perspective is in keeping with Henri Bergson's contention that "time is at first identical with the continuity of our inner life."[4] To be human is to be self-conscious, and therein lies the primordial feeling of duration. Put differently, human beings are aware of their own endurance, and this reflexivity gives human existence an intrinsically temporal character.[5] If, however, temporality is a facet of subjectivity, then it follows that one's sense of duration is shaped from the very outset by society because self-consciousness is generated through socialization.[6]

It was, of course, Emile Durkheim who recognized that the individual's temporal experience is conditioned by the collective rhythms of society. Given his functional orientation, he emphasized that a working consensus on temporality is requisite for the maintenance of social order because consensus is constitutive of intersubjectivity and interpersonal coordination. Indeed, his macrosociological outlook is explicit where he concludes "that a common time is agreed upon, which everybody conceives in the same fashion."[7] Here, Durkheim let the matter rest, but two important qualifications must be noted.

First, socialization notwithstanding, human beings still experience duration in a seemingly idiosyncratic manner. A minute may be a minute, but is there anyone who has not had the sensation that time has passed quickly or slowly? Clocks and calendars represent cultural unanimity, and they have made for some standardization of temporality.[8] Near the turn of the century, for example, Georg Simmel asserted that metropolitan "precision has been brought about through the general diffusion of pocket watches."[9] Nonetheless, standardization and socialization do not prevent individuals from intermittently experiencing personal disjuncture with the time of clocks and calendars. One could argue, in fact, that the experience of personal disjuncture is made more visible by viewing it against the backdrop of temporal orthodoxy.

Second, social reality is not the monolithic entity that Durkheim makes it out to be. In his seminal essay, "The Perception of Reality," William James describes several "worlds" or "sub-universes" of social reality.[10] He observes that the world of science is not the same as the world of religion, and that the world of religion is not the same as the world of common sense. This way of thinking especially influenced the work of Alfred Schutz, whose essay, "On Multiple Realities," picks up where the writings of James leave off.[11] For Schutz, social reality is partitioned into "finite provinces of meaning," each of which is characterized by "a specific time-perspective."[12] For instance, divergent forms of temporal experience help to distinguish dreams from the wide-awake world of everyday life. In turn, Erving Goffman was influenced by both James and Schutz when he wrote *Frame Analysis: An Essay on the Organization of Experience*.[13] Goffman argues that one does not know how to act until one has defined, or "framed," the situation at hand. By classifying situations in this fashion, human beings define a multiplicity of realities, each of which is invested with its own kind of meaning—such as a trial, versus a mock trial in law school, versus the cinematic portrayal of a mock trial in law school.

Various students of temporality have asserted that lived duration assumes the appearance of idiosyncrasy when, in actuality, it is structured by the pluralistic properties of social reality. Thus, Georges Gurvitch describes "multiple manifestations of time," and Edward Hall points out that temporal experience is "situation- as well as culture-dependent."[14] In contrast to these static formulations, Alfred Schutz and Thomas Luckmann emphasize the dynamic relationship between lived duration and the process of social interaction:

The temporal articulation of the stream of consciousness is determined by the tension of consciousness, which alters with transitions from one province of reality with finite meaning-structure to another, as well as, to a lesser extent, with transitions from one situation to another within the everyday life-world.[15]

Yes, but *how* does this transpire? We go from work to play, from civility to violence, from indifference to religious or political fervor, from dreams to the wide-awake stance of public accessibility. Our experience of duration is thought to change as we step from one realm of social reality to another, or as the immediate situation mutates into a new form of interaction. Obviously, Schutz and Luckmann's statement does not take us very far, but little has been done to extend our understanding beyond their promising insight.

In this book, I address the following question: How is the perceived passage of time shaped by the individual's transition from one to another situation or realm of social reality? This question implicates the interplay of subjectivity and objectivity in temporal experience. My response to this question is based upon the theoretical traditions of symbolic interactionism, microstructuralism, phenomenology, and ethnomethodology. In particular, I will be concerned with a "phenomenological analysis of time-consciousness."[16] I want to know how processes associated with self and situation condition the perceived passage of time. Such knowledge will not only advance the study of temporality and social psychology, but will also help us understand often uncanny experiences in everyday life.

The Backdrop: Temporality and Society

Ulysses, by James Joyce, is 783 pages in length.[17] Yet that book recounts only one day in the lives of its two primary characters. How much human experience can fit within a single day? How much in a single hour, or, for that matter, in any standard unit of temporality? For now, it may be impossible to answer these questions with any specificity, but this much is certain: there is fluctuation in the density of experience (i.e., conscious information processing) per standard temporal unit, and that fluctuation is governed by the interplay of self and situation. There is, nonetheless, a tendency for sociological students of temporality to gloss over the basic fact that what feels like minutes for one person may feel like hours for another. This offhand treatment is remarkable because even the most cursory reflection indicates that variation in the experience of time does not occur because there are different kinds of people but because people find themselves in different kinds of circumstances.

One might think that, with his penchant for the study of mind and self, this topic would have been of interest to George Herbert Mead.[18] Such is not the case. In fact, Mead is mute on the subject of duration, but he has much to say about how temporality figures in the dynamics of self-consciousness and social interaction. Mead's interest in temporality stems from a desire to use evolutionary principles in an analysis of mind, with the latter viewed as an ongoing adaptation to changing circumstances.[19] Consequently, Mead has to account for the emergence of new forms of consciousness. As he puts it, the "novelty of every future demands a novel past."[20] Mead fundamentally alters the stimulus-response arc of behaviorism by insisting that, in human interaction, there is an intervening moment—the "specious present"—during which the individual interprets the situation and self-consciously considers alternative lines of action.[21] Furthermore, he emphasizes that the resulting "response . . . is something that is more or less uncertain."[22] Mead is intent on showing how human beings redefine the past from the standpoint of the present, and how such redefinitions create the possibility of novel futures.[23] In this manner, he tries to reconcile determinism and emergence.

It is unlikely that Mead would have endorsed research on a possible link between the volume of experience and perception of the passage of time, although such research is implied by his own tenets. He contends that "the unit of existence in human experience is the act."[24] According to Mead, the act is composed of four phases: impulse, perception, manipulation, and consummation.[25] As but one phase among four, perception does not warrant separate attention beyond the part it plays in the social act. And, from Mead's standpoint, experience is embedded within the social act because experience proceeds through self-reflection: "Action of the organism with reference to itself is, then, a precondition of the appearance of an object in its experience."[26] Moreover, Mead tells us that experience can only be grasped from a particular social perspective: "experience . . . has its character over against actual or possible audiences or observers whose selves are essential to the existence of our own selves."[27]

Mead's subordination of experience to action does not square with his own argument that action is prefigured in self-consciousness as one considers various hypothetical solutions to the problem at hand. Perhaps his pragmatism is showing. A particular aspect of experience may or may not lead to consummation of the act; it may or may not have practical (i.e., behavioral) effects. Nevertheless, there is more to life than solving problems, and, with Norman Denzin, I would assert that Mead's position unnecessarily restricts the symbolic interactionist enterprise:

By placing priority on the act, Mead shifted attention off the ongoing flow of temporal experience that shapes the context wherein the act supposedly occurs. For him . . . the act takes precedence over the moment. . . . *The act, not temporal experience, thus becomes the key concept for Mead, and by implication the generations of interactionists who have followed him.*[28]

If we change the emphasis from action to consciousness, then we can examine a crucial facet of human experience—

variation in the perceived passage of time—that has no place in Mead's theoretical framework.[29]

"There is a certain temporal process going on in experience," states Mead.[30] I will show that there is also a certain experiential process going on in temporality. This change in emphasis implicates the writings of phenomenologists.[31] During the last years of Mead's life (he died in 1931), European phenomenologists were taking up the topic of temporality in more direct fashion than he did. For example, Martin Heidegger is critical of the way ordinary speech depicts time as if it were autonomous from human existence, and he views temporality as an "integral part of human experience."[32] It follows that variation in temporal experience reflects variation in social conditions. Thus, Heidegger proposes that "time . . . functions as a criterion for distinguishing realms of Being."[33] This proposition mandates the investigation of subjective and situated processes that shape the perceived passage of time.

Consciousness is composed of successive experiences: "By thinking, we move through time."[34] But how is it that we perceive duration in successive experiences? Edmund Husserl suggests that the essence of duration can be found in the interplay of "memory and expectation."[35] He points out that we only ever hear one note at a time when listening to a song. Our grasp of the melody results from our ability to remember previous notes and anticipate those which are yet to be played. Husserl differentiates memory into primary remembrance or "retention" and secondary remembrance or "recollections." Each particular note of the melody is perceived now, that is, in the present, but Husserl observes that each moment perceived as the present draws behind itself "a comet's tail of retentions."[36] This is in contrast to our recollection of those past events that are not immediately related to the present. Similarly, he differentiates anticipation into primary expectation (or "protention") and secondary expectation. Husserl concludes that human beings have a "temporally constitutive consciousness" which produces the experience of duration by integrating perception, memory, and anticipation.[37]

Eugène Minkowski provides us with the concept of "lived time," by which he refers to the experience of duration.[38] His primary interests concern disjunctures between lived time and the time of clocks and calendars. He is especially attuned to such disjunctures as a consequence of his inquiries into the temporal anomalies that accompany various forms of psychopathology. Like Heidegger, Minkowski sees temporality as inextricably enmeshed in human existence, and this includes "our states of consciousness, as well as the events which unfold around us."[39] Not only is activity "a phenomenon of temporal nature," but the emotions are also structured by temporality.[40] We fear that which has not yet happened, but we are angered by what has already transpired. Memory encompasses pride or remorse, while anticipation spawns hope and desire as well as dread. The blossoming of our conduct and consciousness thrusts us into the future. Being is becoming, and, for Minkowski, this means that time is a "synonym of life in the broadest sense of the word."[41] His pathbreaking studies foreshadow a basic principle: to experience the passage of time is to make something of one's immediate circumstances.

In contrast to the phenomenological exploration of subjectivity, sociologists emphasize the social organization of temporality as manifested in clocks, calendars, schedules, seasons, and other culturally defined periodicities. They note that members of our society are socialized to an intersubjective structure of standard temporal units, such as years, months, weeks, days, hours, minutes, and seconds.[42] They depict time as a scarce resource and a fundamental aspect of interpersonal coordination.[43] These studies illuminate behavioral patterns and collective representations, but they overlook the fact that temporal organization is not synonymous with the lived experience of time. Research on temporal order points to the need for the development of a social psychology of temporality that embraces the subjectivity as well as the objectivity of time.

The most extensive and influential work in the sociology of time is that of Eviatar Zerubavel.[44] In unequivocal fashion, his prolific publications document the social construction of temporality. He examines the standardization of time as well as efforts to overthrow existing standards. He reveals temporal patterns in social organization, the importance of seasonal holidays for group identity, and the utterly artificial origins of the week. There is, however, an ironic quality to the trajectory of Zerubavel's writings. While his work begins with a Durkheimian emphasis on temporal standardization and the social construction of time, his most recent efforts involve the formulation of a cognitive sociology and the exploration of subjectivity. Through his "phenomenology of the social world," Zerubavel shows us how to bridge the gulf between a macrosociological concern with structure and a microsociological concern with experience.[45]

The potential for research on the subjective aspects of temporality is evident in the symbolic interactionist literature on illness and dying. Early investigations revealed that patients interpret time spent in a hospital by reference to "time perspectives" or "timetable norms" which enable patients to anticipate how long their treatment will take.[46] Similarly, the staff at a hospital comes to anticipate certain "dying trajectories" on the basis of their "experiential careers."[47] More recently, David Maines has examined the "diabetic experience as lived time," and Helena Lopata has described the altered temporal experience occasioned by widowhood.[48] Kathy Charmaz summarizes her own findings, as well as those of her colleagues, when she states that "being ill gives rise to ways—often new ways—of experiencing time."[49]

These studies provide a wealth of insight concerning the particular temporality of illness and dying, but the findings do not "add up" to a general theoretical understanding of variation in the perceived passage of time. The symbolic interactionist paradigm does not exclude scientific generalization. Indeed, Carl Couch calls for research that is directed toward the isolation of "generic sociological principles."[50] My interests are social psychological rather than strictly sociological; nevertheless, I am intent on the discovery of those principles that make for "generic temporal processes."[51] With respect to the experience of duration, such principles, taken together, would constitute systematic specification of the formal or generic properties of lived time. In this book, I formulate a general theory of lived time and examine that theory in light of empirical materials.

The Backdrop: Temporality and Psyche

Sociological attention to variation in the perceived passage of time has been modest and intermittent. This is not true of psychology, where our colleagues have produced a vast research literature on temporal experience.[52] Despite its size and long history, however, Joseph McGrath and Janice Kelly describe this literature as "incomplete and containing unresolved issues."[53] Indeed, the leading theoretical models are contradictory as well as unpersuasive, and they leave important aspects of the phenomenon unexplained. In this section, I review the leading theoretical models and examine some of their unresolved issues.

The Linear Models

Psychologists used linear models in their earliest efforts to conceptualize variation in the perceived passage of time. William James conceived one of the first of these models: "*In general, a time filled with varied and interesting experiences seems short in passing, but long as we look back. On the other hand, a tract of time empty of experiences seems long in passing, but in retrospect short.*"[54] In this model, the perceived passage of time is presumed to be a linear function of the eventfulness that characterizes a given interval. However, while James's formulation has the virtue of incorporating the element of memory, it does not encompass much of the variation in the perceived passage of time. In fact, under certain conditions, a "busy" interval is experienced as passing very slowly. Consider what happens to the experience of time when an undercover police officer finds himself in an ambush:

I go into the stupid buy and the shit starts right from the get-go. We start up the stairs and George tells me it's the fourth floor; then he changes it to the fifth. As we're arguing and walking up, he keeps getting in back of me and lagging. I'm grabbing him and pushing him in front of me. The fuck don't want to go up. It was only about two floors up and we got to a landing and some fucking guy comes out with a .45. It was a setup. The fucking guy, George, set me up. I couldn't believe it.

For some shitass reason I had this little .25 automatic hidden down in this like jockstrap. All I remember is trying to get into my fucking pants and he was shooting. That was a very bad situation that seems to be going by very slowly, when in fact it all happened within a matter of seconds.[55]

Obviously, this encounter is "filled with varied and interesting experiences," but, James to the contrary, the individual in question reports the perception that time is passing slowly.

In addition, this formulation contradicts Robert Ornstein's "cognitive, information-processing approach" to conceptualizing the experience of time.[56] Like James, Ornstein develops a linear model in which memory is an important component, but, unlike James, Ornstein argues for a positive relationship between the complexity of an episode and one's judgment of the passage of time: "We would expect that an increase in the number of events occurring within a given interval, or an increase in the complexity of these events . . . would each lengthen the experience of duration of that interval."[57] Ornstein's formulation fits the fact that time seems to go by slowly during episodes of great complexity—for example, the ambush reported above. However, it runs counter to the commonplace observation that busy time seems to pass quickly while empty time seems to pass slowly. James Taylor, Louis Zurcher, and William Key found that, in the aftermath of a tornado, many of the survivors felt mired in empty time because of their paralyzing concern for the fate of others, and because the bases for their usual forms of activity had been disrupted:

My most immediate reaction was worrying about my friends. The street that seemed to be the worst was right around where they lived. . . . Through all that day, which seemed like one of the longest days of my life, I sat around waiting for news at the office; waiting and wondering if there was anything I could do.[58]

Waiting is one of the classic examples of empty time, yet, in stark contrast to what Ornstein would predict, the survivors experienced time passing very slowly.

Both linear models leave whole categories of the phenomenon unaccounted for. This problem may stem from the way both models emphasize the objective qualities of the situation, or what Ornstein calls "external information," while neglecting the subjective experiences of the people who are caught up in those situations.[59] The moments during the ambush were not only full of overt events, but also covert events: the vivid perceptions, thoughts, and emotions of a police officer who is desperately trying to make sense of what is happening to him. Similarly, the survivor of the tornado was not in fact experiencing "empty" time at all, for he had himself filled that interval with fervent concern for loved ones and intense consideration of what he might do to help them. Thus, the shortcomings of the linear models suggest that the subjective aspects of the situation may be just as important as its objective aspects in shaping variation in the perceived passage of time.

A U-Shaped Model

H. Wayne Hogan contends that we can reconcile the two linear models by combining them in a single curvilinear image. He depicts variation in the experience of time as a U-shaped relationship between the stimulus complexity of the situation and one's inner sense of duration. From this perspective, both linear models are correct but incomplete, as each expresses only part of the relationship. James's argument for an inverse relationship is represented by the left half of the U-shaped curve, while Ornstein's argument for a positive relationship is represented by the right half of the U-shaped curve. By way of explanation, Hogan begins with the familiar idea that "lack of stimulation" is "boring," but he goes on to claim that "boredom also follows from the experience of being subjected to more stimulation (i.e., to sensory overload) than the individual's experiential system can accommodate."[60] This leads Hogan to conclude that the passage of time is perceived to accelerate during intervals with moderate stimulus complexity: "It therefore remains for *moderately* complex stimuli to be experienced as comparatively, relatively 'fuller,' hence shorter, than either minimally or maximally stimulating time intervals."[61]

Hogan asserts that there is empirical support for his model by using experimental design to show that "stimuli both least and most complex are experienced as involving lengthier durations than stimuli of moderate complexity."[62] Moreover, at first glance, the narratives concerning an ambush and the aftermath of a tornado provide anecdotal corroboration, the former representing situations of great overt complexity, the latter settings where overt complexity is lacking. Hogan wants to grasp the common element in these seemingly different settings, but are they equally full or equally empty? He elects to argue that they are equally empty. This decision enables him to integrate opposing models, but it also forces him to resort to a convoluted and questionable logic. Already, it is apparent that so-called "empty" intervals are actually quite full when one considers the subjective processes of the people who are in those situations. By arguing that the settings are equally empty, Hogan must make the dubious contention that settings of high stimulus complexity are just as "boring" as settings of low stimulus complexity. Undoubtedly, the police officer was experiencing a number of emotions during the ambush, but his narrative makes it plain that boredom was not among them. Likewise, Hogan is compelled to argue that intervals defined as having moderate stimulus complexity are experienced as "fuller" than intervals defined as having high stimulus complexity.

Another problem with Hogan's formulation is that it does not comprehend the full range of variation in the perceived passage of time. This stems from the research method he adopts from his predecessor. Like Ornstein, Hogan does not compare subjective time to the intersubjective time of clocks and calendars.[63] The individuals in his laboratory are asked to indicate whether one interval is "shorter than, equal to, or longer than" another interval of the same objective length, but containing a different amount of stimulus complexity. Any one interval, then, can be experienced as relatively shorter or longer than another, but an interval cannot be experienced as absolutely short or long by contrasting one's inner sense of duration with the measurement of that interval in terms of standard temporal units, such as hours, minutes, and seconds. This is in keeping with Ornstein's fundamental presupposition: "An analysis should be concerned with experiential time per se, not as it might relate to hours, days, burning rope or to some other time definition."[64] Thus, both scholars favor laboratory methods, thereby abstaining from research on the experience of time in everyday life.

Yet, as is evident in the narratives above, people in real life commonly translate anomalous temporal experiences into standard temporal units. This practice reflects the crucial role of clocks and calendars in the structuring of social interaction. If, as Hogan argues, the passage of time is perceived to slow under conditions of high or low stimulus complexity, and if it is perceived to quicken under conditions of moderate stimulus complexity, then under what conditions is our inner sense of duration roughly synchronized with the intersubjective time of clocks and calendars? Ordinarily, the individual does not experience time passing quickly or slowly; instead, ten minutes of time as measured by the clock *feels* like ten minutes from the subjective standpoint of inner duration. That this is the typical experience of time in everyday life should come as no surprise because an approximate synchronicity is necessary for the sake of mutual predictability and the coordination of one's conduct with others. A theoretical model of variation in the experience of time must account for the possibility—indeed, the prevalence—of synchronicity.

Despite its serious shortcomings, Hogan's U-shaped model does demonstrate the importance of attending to the objective amount of stimulus complexity that characterizes the setting. Furthermore, Hogan improves on the earlier work of James and Ornstein by recognizing that time is perceived to pass slowly in situations of high *or* low stimulus complexity. An adequate theoretical model must explain the underlying commonality that produces the same inner sense of duration under these seemingly different circumstances.

Plan of the Book

How can we account for variation in the perceived passage of time? Despite considerable effort, this question remains unresolved. The persistence of this problem does not stem from neglect so much as reluctance to pursue research within the empirical confines of everyday life. By and large, scholars have emphasized experimental control at the expense of thorough familiarity with naturally occurring variation. In chapter 2, I take the measure of that variation by establishing what must be explicated: the three paradoxical aspects and three elementary forms of lived duration. In chapter 3, I present a great deal of qualitative data to acquaint the reader with situations in which time is perceived to pass slowly. In chapter 4, I construct a theory that is capable of accounting for the full range of variation in the perceived passage of time. In chapter 5, I derive hypotheses from that theory concerning factors that make for the perception that time has passed quickly, and I examine those hypotheses in light of qualitative and quantitative data. In chapter 6, I address the theoretical implications of this study and consider directions for future research. In the Methodological Appendix, I discuss the principles that guided the collection and analysis of my data.

[2]

Paradoxical Variation

OW DO WE experience variation in the perceived passage of time? Logic dictates that we address this question before we ask why it is perceived in that fashion. King, Keohane, and Verba provide us with a rationale: "It is pointless to seek to explain what we have not described with a reasonable degree of precision."[1] This is sensible advice, but, as we have seen, it did not always guide the work of earlier scholars. Many of the problems with the ways they conceived of lived duration are derived from their failure to recognize its full range of variation. Admittedly, this is no easy task. The conscientious student of lived duration must somehow come to grips with a facet of subjectivity that is paradoxical, variegated, and taken for granted by laypeople and scientists alike.

In this chapter, I attempt to describe the full range of variation in the perceived passage of time. It is my intention to map the contours of lived duration, thereby marking the perimeter of that territory which is explored more thoroughly in the chapters that follow this one. Consequently, only enough of the data are introduced here to establish the extent of variation, not to account for it. In the first section, I discuss three paradoxical aspects of lived duration, and in the second section, three elementary forms of lived duration. Together, these six facets of variation in the perceived passage of time represent the experiential possibilities that must be accommodated by a comprehensive theory of lived duration.

Three Paradoxical Aspects

The first paradox of lived duration is that time is perceived to pass slowly in situations with abnormally high *or* abnormally low levels of overt activity. There is a folk theory that claims busy time goes by quickly while empty time goes by slowly. This folk theory does not fare well, however, when it is examined in light of systematic observation. It quickly becomes apparent that objectively eventful or objectively uneventful circumstances can provoke the feeling of interminable duration. Witness, for example, the following narrative, in which John Van Maanen gives a first-person account of his participation in the pursuit of a stolen car by the police:

Except for the sergeant still sitting in the Mercedes, the accident scene empties quickly. I start the car and drive away, slowly realizing that I don't have a clue how to handle this mobile example of police high tech.

The radio seems to be screeching at me to do something. The lights and siren, to my astonishment, somehow come on. The demonic shotgun is no longer secure and bounces around the front seat. The power brakes feel awkward and almost toss me through the windscreen at the first stop sign. To complicate matters, I have no idea where I'm going.

As I round a corner near the Interstate, the ticket book, the clipboard, the logbook, the portable radio, David's hat, and God knows what else go sliding out the passenger door I'd forgotten to fully close and onto the street. The shotgun would have gone too had I not grasped the stock of the weapon with a last-second, panic-stricken lunge. Shamefully, I pull to the side of the road to gather up my litter.

Luckily I was not unobserved by the police. Two Southend officers, looking for something to do, had come up to assist in the chase. . . .

They drive up as I'm puzzling over what switch controls what function of the machine. . . . Between chuckles, they give me remedial instructions on how to operate a prowl car and direct me to where I am next to appear.

When I eventually arrive at the firehouse, David is standing on the corner chatting with another officer. Maybe ten minutes have passed since I left the section sergeant fondling the Mercedes, but of course it seems like years. William James is right about time stretching out when events conspire to fill it up.[2]

Van Maanen provides us with detailed description of a busy interval that evokes within him the feeling that time is passing very slowly. Moreover, his observation corresponds with those of hundreds of others who found time to pass slowly in combat, automobile accidents, and other highly eventful circumstances. However, if "William James is right about time stretching out when events conspire to fill it up," then why is there a folk theory that tells us *empty* time passes slowly? And, in fact, it is easy to find evidence that supports the folk theory, evidence from some of the most familiar forms of frustration in everyday life. The excerpts below are taken from interviews with three students. To begin with, someone at the other end of a telephone line may put us "on hold":

I dialed the office and his secretary answered. I asked if I could talk to the lawyer and she transferred me to another secretary. Again, I asked to talk to the lawyer, and she remarked, "Please hold." So I held, and it felt as if it was forever, making me feel frustrated and angry.

At other times, we may be left to cool our heels in a doctor's waiting room:

The waiting room . . . was crowded so I knew it would take a while. The receptionists seemed disorganized and inefficient. I wanted my visit to be brief, but felt no one else cared about the plans I had for traveling that morning. I waited again in the doctor's examining room until I was actually examined. . . . The time seemed to go by extremely slowly.

Or, we may have to endure an uneventful evening at work:

The day started slowly and seldom picked up in business. I usually have to open up a register during the day, but didn't that day. People seemed to be preoccupied and didn't socialize much. Others were coming and going while I stayed and watched the exchange. Time felt as if it was creeping along. Looking up at the clock seemed to be worthless because the hands moved so slowly. When the end of the night came . . . I felt as if I had been there forever.

And, by means of more intrusive inquiry, the absence of overt activity can be pushed to exotic and technological extremes, albeit with the same effect on lived duration. In the following narrative, a male, forty-year-old physics professor reports on his experiences inside a stimulus deprivation tank:

I was instructed (before entering) to try to stay as long as I felt comfortable and to try to keep track of time. . . . It seemed pleasant and relaxing—an escape. Time seemed to pass slowly. The longer I stayed, the more slowly it seemed to pass.

These stories demonstrate that protracted duration cannot be reduced to "time stretching out when events conspire to fill it up." Whether the setting is found in everyday life or the more contrived context of a laboratory, the perception that time is passing slowly is much more complex and ambiguous than anything envisioned in the superficial analysis of William James. What is more, as the last narrative makes clear, anger and frustration in response to boredom are not necessary aspects of this experience. In short, time is perceived to pass slowly in situations of abnormally high or abnormally low overt activity. A comprehensive theory of lived duration must resolve this paradox.

The second paradox of lived duration is that the same interval of time which is experienced as passing slowly in the present can be remembered as having passed quickly in retrospect. Again, James was well aware of this enigma, as is anyone who, like Albert Camus, ponders the relationship between imprisonment and lived duration:

I'd read, of course, that in jail one ends up by losing track of time. But this had never meant anything definite to me. I hadn't grasped how days could be at once long and short. Long, no doubt, as periods to live through, but so distended that they ended up by overlapping each other.[3]

Here, Camus alerts us to the interplay of perception and memory. The perceived passage of time is conditioned by differentiation between the present and the past. In other words, an interval of time in the immediacy of one's current experience will not be identical to that same interval of time once it has been consigned to one's memory. Moreover, the statement by Camus implies that the interplay of perception and memory is conditioned further by what transpires during a given interval. Thus, the emptiness and paralysis of imprisonment will not be remembered in the same way as another interval of equal objective length, but one which is filled with variety and activity.

Both of these points are made with greater clarity and specificity in a narrative taken from Arthur Koestler's book, *Dialogue with Death*. In this memoir, he describes the ordeal of solitary confinement subsequent to being captured by loyalist forces in the Spanish Civil War. During the course of his imprisonment, he becomes obsessed with the elasticity of temporal experience:

The astonishing thing, the puzzling thing, the consoling thing about this time was that it passed. I am speaking the plain unvarnished truth when I say that I did not know how. I tried to catch it in the act. I lay in wait for it, I riveted my eyes on the second hand of my watch, resolved to think of nothing else but pure time. I held it like the simpleton in the fable who thought that to catch a bird you had to put salt on its tail. I stared at the second hand for minutes on end, for quarters of an hour on end, until my eyes watered with the effort of concentration and a kind of trance-like stupor set in. . . .

Time crawled through this desert of uneventfulness as though lame in both feet. I have said that the astonishing and consoling thing was that in this pitiable state it should pass at all. But there was something that was more astonishing, that positively bordered on the miraculous, and that was that this time, these interminable hours, days and weeks, passed *more swiftly* than a period of times [*sic*] has ever passed for me before. . . .

I was conscious of this paradox whenever I scratched a fresh mark on the white plaster of the wall, and with a particular shock of astonishment when I drew a circle round the marks to celebrate the passage of the weeks and, later, the months. What, another week, a whole month, a whole quarter of a year! Didn't it seem only like yesterday that this cell door had banged to behind me for the first time?[4]

Through the ordeal of imprisonment, Koestler stumbles upon one of the three paradoxical features of lived duration. From moment to moment (i.e., current experience), his time in solitary confinement seems to pass slowly, but, upon reflection (i.e., with hindsight), it seems to have passed quickly. Furthermore, both of these feelings provide implicit contrast with the normal, nearly unnoticed synchronization between lived duration and the passage of standard temporal units. There is, then, this puzzling elasticity to the perceived passage of time.

Koestler is struck by what he calls the "uneventfulness" of his time in solitary confinement, but it would not be accurate to say that this interval is "empty," as Hogan would have it.[5] On the contrary, it is quite apparent that Koestler himself fills his time in prison with attention to all manner of subjective and objective experiences. He stares at his watch; he thinks about time; he makes chalk marks on the walls of his cell. And, of course, there were any number of other things that do not appear in his narrative—things that were thought, felt, and done in a desperate attempt to while away the long months of his solitary confinement. It is in this sense that "just killing time" can be a very active endeavor.

Some inkling of the resourcefulness he displays in dealing with his predicament is evident in the following passage, where he reveals rather more knowledge of his physical setting than is typical of those who are free to come and go as they please: "I racked my brain in an effort to explain to myself these paradoxes of time, while I paced up and down between my bunk and the W.C.—six and a half paces there, six and a half paces back."[6]

Koestler describes his imprisonment as "the most absolute degree of uneventfulness imaginable."[7] However, it should be obvious that this claim is belied by the thoroughness with which he relates the details of his experience. Koestler's own words show that his time in solitary confinement was not the utter vacuum he makes it out to be. It is true that no dramatic or memorable events punctured the monotony of each passing day, but he is overlooking the success of his own valiant effort to substitute attention and concentration for intrusion and distraction.

Likewise, consider what transpires when a young woman arrives at an office for her first day as a lawn-care worker:

At exactly eight, I walked through the door. There were approximately twenty people in the room, all different ages and sizes. Each person (except for the secretary) was wearing a yellow shirt, neon green pants, and rubber boots. I was wearing a different color yellow shirt and dark green pants. I also had on hiking boots. When I walked in, every pair of eyes in the room was staring at me. They were checking me out and judging me. I was taller than all of the women and quite a few of the men. I was also the only blond. No one said a word, so neither did I. I stood in the doorway waiting for someone to tell me what to do, but no one did. At this point I was feeling very embarrassed and vulnerable. I was wishing I had never shown up. I felt as though I had been standing there for hours. I had no one there to talk to. Time passed very slowly so it seemed, but in reality, it had only been two minutes.

A detached observer might note that no one is "doing" anything. Yet it is obvious that our narrator is utterly absorbed by a heightened sensitivity to self and surroundings. The intensity of her attention to the internal and external environments of mind *creates* stimulus complexity in the midst of a situation that is all but devoid of overt activity.

The more closely one looks at episodes of "empty" time, the less empty they seem to be. Indeed, under more careful scrutiny, they turn out to be quite full—not of demonstrable events or activity in an objective sense, but of perceptions, feelings, thoughts; in a word, of *engrossment* with the proceedings, such as they are. Ironically, when we asked respondents about uneventful episodes, they reported that "nothing" was happening even though, like Koestler, their own descriptions contradicted this conclusion. In fact, a great many things are happening as the young woman waits in that office and—again, like Koestler—much of what she attends to is of her own making. In other words, "empty" time is a misnomer, but not because a passive subject is pummeled by overt, external stimuli. Rather, that which is mistaken for time free of stimulus complexity or embedded activity is actually filled by cognitive and affective processes on the part of those who are covertly, actively, and self-consciously engrossed by their circumstances, regardless of the apparent uneventfulness.

The fullness of "empty" intervals will be an essential element of theory construction in chapter 4. For now, however, we are left with the paradoxical fact that a single period of time can be experienced as passing slowly in the immediacy of the present, yet also be remembered as having passed quickly in retrospect. Our model for variation in the perceived passage of time must reconcile these incongruous aspects of temporality.

The third paradox of lived duration is that some busy intervals are experienced as passing slowly, while others are experienced as having passed quickly. Put differently, situations with abnormally high levels of overt activity seem to result in people perceiving the rate at which time passes in two diametrically opposed ways. Why is it that the duration of a busy interval is sometimes perceived as longer, and at other times shorter, than its actual length (as measured by a clock or calendar)?

Once again, common-sense knowledge does not rescue us from our dilemma. The folk theory tells us that busy time passes quickly, and, coupled with our own experiences, the familiarity of this aphorism may make it seem indisputable.

Nevertheless, the folk theory only serves to mark one side of the apparent contradiction. As we have seen, Van Maanen perceived time to pass slowly during a very eventful episode. His account is corroborated by Ansel Adams when the latter finds himself in the midst of an earthquake:

At five-fifteen the next morning, we were awakened by a tremendous noise. Our beds were moving violently about. Nelly held frantically onto mine, as together we crashed back and forth against the walls. Our west window gave way in a shower of glass, and the handsome brick chimney passed by the north window, slicing through the greenhouse my father had just completed. The roaring, swaying, moving, and grinding continued for what seemed like a long time; it actually took less than a minute.[8]

Whatever idiosyncrasies distinguish their respective circumstances, both Van Maanen and Adams are forced to endure an interval of time that is filled with unfamiliar events and vivid experiences. Moreover, despite the seemingly incontrovertible adage, "busy time passes quickly," and in contradiction to the equally simplistic analysis of William James, both Van Maanen and Adams report that time was perceived to pass slowly. Indeed, a given period of time can be so supercharged with harrowing eventfulness that one is left incredulous at how much can transpire within a relatively brief interval. Witness, for instance, Elie Wiesel's shock as he reflects on what his family suffers when soldiers seize them during the Holocaust:

So much had happened within such a few hours that I had lost all sense of time. When had we left our houses? And the ghetto? And the train? Was it only a week? One night—*one single night*?[9]

One might object that the foregoing examples are unrepresentative of what we experience in less exotic settings. This is not the case. Systematic observation demonstrates that the exotic properties of earlier specimens only serve to highlight a relationship that is prevalent, if less obvious, in the circumstances of everyday life. In the following narratives, two young women recount how unusually busy shifts at work affected their perception of the passage of time:

I was checking out food at Albertson's grocery store. I worked for about nine hours, but it felt like days had gone by. The store was very *busy* and I was constantly scanning food and bagging groceries. Time seemed to go by so fast, but only minutes had. Scanning the groceries seemed to take forever, and every time I looked at my watch only minutes had passed. [italics added]

I started at 5:00 p.m. A lot of people surrounded me with various questions, check approvals, and refunds. When I looked at the clock it was 5:15. Yet, it seemed as though I had been there for an hour. More people arrived at six, but the time still went slowly. I felt as though I was going to be stuck in K-Mart forever. Finally, after a *busy* night, it was over. The clock said 9:00 but to me it should have been the next day. [italics added]

Thus, we do not need recourse to earthquakes and pogroms. Eventfulness can have a similarly anomalous effect on the perceived passage of time even when that eventfulness takes the form of nothing more exotic than an unusually busy shift at work. Note that their time at work seems to pass slowly even though the two women in the preceding narratives must deal with abnormally abundant demands on their attention. Earlier in this chapter, we saw that a slow night at work can seem interminable. And, according to folk wisdom, a busy night at work should be experienced as having passed quickly. Now we see that a night at work can be too busy as well as not busy enough, and that either way time is perceived to pass slowly. So, we are confronted with a third paradox: some busy intervals are perceived to pass slowly, while other busy intervals are perceived to pass quickly. Furthermore, Isaac Asimov adds an important addendum by suggesting that suffering is not a necessary component of those busy intervals during which one perceives time to pass slowly:

It was, looking back at it, an idyllic period that followed. A hundred things took place in those physioweeks, and all confused itself inextricably in Harlan's memory, later, making the period seem to have lasted much longer than it did.[10]

It is reasonable to assume that some as yet unspecified factor differentiates those busy intervals which are perceived to pass slowly from those which are perceived to pass quickly. In other words, we must be alert to the possibility that there are, in fact, two kinds of busy intervals. Elaboration on this point must be postponed, however, until we make the transition from description to analysis.

By way of summary, then, anyone who wishes to account for variation in the perceived passage of time must be prepared to resolve three paradoxical aspects of lived duration. First, time is perceived to pass slowly in situations with abnormally high or abnormally low levels of overt activity. Second, the same interval of time which is experienced as passing slowly in the present can be remembered as having passed quickly in retrospect. Third, some busy intervals are experienced as passing slowly, while other busy intervals are experienced as having passed quickly.

Three Elementary Forms

The allusion to Emile Durkheim is intentional.[11] In this section, I discuss the three elementary forms of variation in the perceived passage of time. As with the three paradoxical aspects, the goal here is to ascertain only the outline of the full extent of variation, leaving theory construction for later chapters. Moreover, I shall forego furnishing this section with concrete examples, and rely instead on the reader's capacity to resonate with what are, after all, experiences each of us has had at one point or another.

The elementary forms of lived duration are "elementary" only in the sense that they are indispensable to the existence of variation in the perceived passage of time. Indeed, it is impossible to conceive of lived duration without reference to such variation. For if there is variation in the perceived passage of time, then the variation reflects the fact that, under certain conditions, time is perceived to pass slowly while, under different conditions, it is perceived to have passed quickly. In addition, there is the further implication that, when time is not perceived to be passing quickly or slowly—that is, *abnormally*—lived duration goes largely unnoticed, as it is roughly synchronized with the intersubjective time of clocks and calendars. Therefore, we can distinguish three points along the continuum that represents variation in the perceived passage of time. Each of these points corresponds to one of the three elementary forms, and, for the sake of convenience, I shall refer to these points by name: protracted duration, temporal compression, and synchronicity.

The phrase "protracted duration" refers to one's experience that time is passing very slowly. This is to say that, in certain situations, it feels as if much more time has elapsed than actually would be measured by a clock or calendar. One of the hallmarks of protracted duration is the way people can be observed struggling to put their anomalous experience into words or translate it into standard temporal units. In somewhat disoriented fashion, they will tell you that a given interval seemed to last "for ages," or that it seemed to go on "forever." Alternatively, they will report that the seconds felt like minutes or that the hours felt like days. We are already acquainted with the other hallmark of protracted duration: the paradoxical fact that this experience is produced by conditions characterized by abnormally high or abnormally low levels of overt activity. In short, the first elementary form of lived duration involves episodes that seem to pass with uncanny slowness.

So, it is apparent that time does not always fly, Virgil's famous dictum to the contrary. Nevertheless, the perception that time has passed quickly is one of the elementary forms of lived duration. I shall refer to it as temporal compression because, under certain conditions, it feels as if much less time has elapsed than actually would be measured by a clock or calendar.[12] While protracted duration and synchronicity are primarily phenomena of the present, temporal compression is a phenomenon that is uniquely associated with the past. Its most familiar manifestation is the shocked look backward that is expressed in common questions: Where have the hours (days, months, or years) gone? Thus, almost all intervals seem to have passed quickly (or compressed) when viewed retrospectively. However, this last statement does not take into consideration what we already know about the third paradoxical aspect of lived duration. To repeat, some busy intervals are experienced as passing slowly, while others are experienced as having passed quickly. This means that temporal compression cannot be reduced to the effects of retrospection.

Ordinarily, it is not our perception that time is passing quickly or slowly, because the perceived passage of time is roughly synchronized with the flow of standard temporal units. Put differently, ten minutes as measured by the clock *feel* like approximately ten minutes from the personal standpoint of one's own subjectivity. I shall refer to this experience as synchronicity, and it is the third elementary form of lived duration.[13] Synchronicity is probably the most prevalent of the elementary forms, and with good reason: it is part of the underpinning for interpersonal coordination and social order. Yet, by virtue of its very prevalence and necessity, synchronicity is a tacit and largely taken-for-granted aspect of social interaction. Consequently, it is the most difficult form of lived duration to study, as people in everyday life do not find it remarkable, and they are puzzled when asked about it. Strictly speaking, then, synchronicity is not quite a facet of consciousness. Ironically, its presence is marked by the near absence of one's attention to the passage of time.

It is worth emphasizing that the elementary forms of lived duration are not discrete categories. On the contrary, the terms which have been introduced in the preceding paragraphs—protracted duration, temporal compression, and synchronicity—refer to points along a continuum that represents variation in the perceived passage of time. There is, of course, convenience in having terms of reference for the three elementary forms of lived duration, but, as Anselm Strauss points out, names have a way of suggesting distinct boundaries between separate entities.[14] With regard to variation in the perceived passage of time, this is not the case; rather, the elementary forms of lived duration should be thought of as tendencies with gradations that shade into one another. What is more, these gradations correspond to modulation in those circumstances that bring about each form of temporal experience. It follows that the elementary forms of lived duration not only differ from each other, but also differ within themselves in terms of intensity. For example, the lived duration of a given interval may seem to pass slowly, or it may seem like it will never end.

It should also be noted that the elementary forms of lived duration represent the real experiences of people in natural settings. This is to say that the elementary forms are phenomena of everyday life, not creatures of the laboratory. Without prodding or manipulation, and in their own words, people tell us about circumstances within which they perceived time to pass slowly or quickly. In addition, these stories provide implicit contrast with their more typical experience: an approximate synchronization between their perception of the passage of time and the actual flow of standard temporal units. Therefore, the elementary forms of lived duration are not by-products of research and reactivity; they are facets of human experience and expression even when inquiry is not at issue.

To sum up, then, there are circumstances in which time seems to pass slowly, circumstances in which time seems to have passed quickly, and circumstances in which one's experience of duration is roughly synchronized with the time of clocks and calendars. A theory that purports to account for variation in the perceived passage of time must show how these circumstances differ from one another, such that they result in the three elementary forms of lived duration.

Folk Wisdom

It is not unusual to hear people describe their experiences in ways that include reference to temporality. However, while temporality can be a remarkable aspect of one's experience, that does not mean it will be viewed as an important aspect of one's experience. Still, most people find temporality interesting—a fact that both helps and hinders research on this topic.

With its three paradoxical aspects and three elementary forms, the student of lived duration confronts a complex and intimidating range of variation. Yet, as is so often the case in the social sciences, those who study lived duration find themselves investigating phenomena that are familiar to people in everyday life. Moreover, familiarity may lead scholars and laypeople alike to think that everything there is to know on this subject is encapsulated within a few simplistic aphorisms. This is a particularly virulent problem in America, where an anti-intellectual tradition exalts folk wisdom far beyond its real worth.

At the outset of this chapter, however, I noted that folk wisdom does not fare well when examined in light of systematic research. Indeed, one of the themes of this chapter has been the rather glaring inadequacies of common-sense knowledge where lived duration is concerned. Folk wisdom tells us that "time flies when you're having fun," but we have seen that pleasurable experiences can seem to transpire very slowly. Folk wisdom tells us that "busy time passes quickly while empty time passes slowly." But we have seen that time can be perceived to pass slowly in situations where there is an abnormally high level of overt activity, while empty intervals typically seem to have passed quickly in retrospect. And, of course, folk wisdom tells us that "time flies," but we have seen that there is a wide variety of circumstances in which time seems to pass slowly. Thus, at every turn, folk wisdom is contradicted by the actual diversity of human experience.

This is not to say that folk wisdom is utterly useless. After all, it represents the distillation of at least some segment of variation in the perceived passage of time, and, consequently, it serves as a helpful source of data. Nonetheless, these aphorisms pass for wisdom only so long as they are kept separate from each other and trotted out at propitious moments. When they are assembled in one place, it quickly becomes evident that they lack consistency (e.g., "time flies" versus "empty time passes slowly") and they lack corroboration (i.e., each of the aphorisms disregards significant segments of human temporal experience). What is worse, these half-truths masquerading as wisdom do not add up to a comprehensive understanding of lived duration. Given stricter standards for consistency, corroboration, and generalization, perhaps the social sciences can succeed where folk wisdom fails.

The theory that emerges from this study is counterintuitive, but thoroughly grounded in a wealth of data, and able to resolve heretofore paradoxical aspects of temporality. Put differently, the findings in this book represent research and discovery, not the reiteration of common sense.

[3]

Protracted Duration

THE VELOCITY AT which one is traveling is most noticeable during periods of acceleration or deceleration. So, for example, we are more aware of the speed of an airplane during takeoff and landing than while it is en route to its destination. Indeed, during the flight, one has to look out a window for confirmation that the plane is even moving, because there is little or no sensation of velocity inside the cabin.

There is, then, a kinship between velocity and duration. The kinship stems from the fact that both concepts have to do with the speed at which something is happening. Under ordinary circumstances, lived duration is a taken-for-granted feature of what Erving Goffman calls "the organization of experience."[1] This means that lived duration becomes a largely tacit aspect of interpersonal coordination whenever the perceived passage of time is roughly synchronized with the intersubjective time of clocks and calendars. People are fully conscious of lived duration only on those occasions when time is perceived to be passing at an abnormal rate (i.e., either too quickly or too slowly). Therefore, given the questions that prompt this study, it was imperative to record descriptions of situations in which people reported having felt the pace of duration alter appreciably.

Extraordinary circumstances make for abnormal temporal experiences, and the latter bring time to the surface of consciousness. Consequently, the stories people tell about lived duration provide only instances of deviant temporality. More precisely, they provide almost nothing but stories about unusual situations in which time was perceived to pass slowly. (People are not apt to tell stories about the perception that time has passed quickly because, as we will see in chapter 5, the circumstances that bring about temporal compression are not memorable.)[2] At first glance, this may seem like an obstacle to inquiry, but, in the manner of Emile Durkheim, we can use this form of deviance to obtain analytical leverage on a facet of experience that is, under ordinary circumstances, difficult if not impossible to study.[3]

From 1983 to 1994, my assistants and I have collected 705 first-person accounts of situations in which the passage of time was perceived to slow noticeably. Following Goffman's lead, 389 of these cases have been "cited from the press and popular books of the biographic genre."[4] The remaining 316 cases have been generated through interviews with students. These data consist of what Eviatar Zerubavel calls "temporal anomalies."[5] His phrase refers to departures from the normative pattern of temporality. Temporal anomalies become manifest whenever any feature of temporal orthodoxy is violated. For the purposes of this chapter, however, the definition of temporal anomaly has been restricted to only one of its various forms: the perception that time is passing slowly, or what I have called protracted duration.

Several considerations recommend this definition. We are told that lived duration changes with transitions from one domain of social reality to another. The foregoing definition allows us to examine that proposition by serving retroactively as an indicator of transformations in the immediate circumstances of social reality. Moreover, in sensitizing us to those contexts that suffice to provoke the experience of protracted duration, this definition enables us to illuminate its etiology. Finally, it is accessible in everyday life; the perception that time is passing slowly is neither so common as to be invisible, nor so rare as to be wholly unfamiliar to people on the street.

My empirical materials are composed of personal testimony that summarizes hundreds of incidents in which time was felt to pass slowly. Another criterion for inclusion was that protracted duration be experienced *during* the episode; narratives were not included if time was perceived to have passed slowly merely as a result of retrospection. If time is said to have passed slowly during all of these incidents, then they can be viewed as representative of those conditions that foster the experience of protracted duration. In the balance of this chapter, I examine these narrative materials in an effort to discover what brings about the perception that time is passing slowly.

Sufficient Causes

The descriptive analysis chronicled in this chapter is directed at theory construction by way of an inductive logic. Careful review of the narrative materials gradually revealed several clear themes with regard to the nature of the circumstances. In descending order of frequency, these themes are (1) suffering and intense emotions, (2) violence and danger, (3) waiting and boredom, (4) altered states, (5) concentration and meditation, and (6) shock and novelty. These themes can be thought of as the sufficient causes of protracted duration. In other words, each of these factors is capable of bringing about the perception that time is passing slowly, but none of them are necessary for the experience of protracted duration. In the following pages, a selection of data is brought forward as a purposeful melody of observations. The chosen vignettes have been selected with an eye toward conceptual elaboration and illustration.

Suffering and Intense Emotions

According to David Karp and William Yoels, "Einstein once commented on the relativity of time by saying that two minutes in an uncomfortable situation seems like two hours."[6] Indeed, any form of suffering is capable of provoking the perception that time is passing slowly. As Kristin Luker notes, this is particularly true when pain is augmented by anguish during an illegal abortion:

At noon we went back [to the abortionist] and he told my husband to wait in the waiting room. I went with him through his office and we went out the back door of his office into the huge room—it was a big empty warehouse. . . . And he said to lie down on the table, and then he said, "Now you have to be very quiet because I can't have anyone in the neighborhood hear you scream." So, with no anesthesia, [and twelve weeks pregnant] I had a D&C. . . . I bit my lip and clenched my fists; fortunately I have no fingernails, or I'm sure they would have gone right through my hands. It seemed like it would never end and it was *terribly* painful. I suppose it took ten minutes, but it felt like two hours.[7]

Several of the people we interviewed told stories about visiting a dentist. It would appear that the common factor is related to the way suffering concentrates one's attention on the here and now of self and situation. This young woman's narrative is typical:

The dentist spent about 25–30 minutes working on me, but it seemed like hours. Every time he dropped the drill he had been using, I thought he was through, but each time he picked up another one.

Similarly, coaches of one sport or another regularly inflict suffering for what they claim is the athlete's own good. Like dentists, coaches may have the best of intentions, but, in any case, their drills have an identical effect on lived duration:

For the first few sprints everything seemed normal, even the five second intervals we were given. Then the sun and fatigue . . . set in. I focused all of my attention on a breathing rhythm and tried to block out the growing agony in my legs. I got very hot very quickly [because] we were in full gear. Now it seemed as if I was sprinting for thirty and forty seconds at a time (really only about six or seven) and resting for two. The intervals were counted off with . . . incredible speed. Around sprint number twenty . . . it got worse. The sprints seemed to last for a minute, each stride taking two or three seconds; it was slow motion until the brief rest period, which flew by.[8]

And, of course, suffering can be inflicted by those who do not have our best interests at heart. Torture is the classic example, as is evident in the following passage where a prisoner of war describes ingenious, albeit cruel treatment at the hands of Japanese soldiers:

We were made to stand at attention, with our arms extended, holding the canteens upside down until all the corn had fallen out. Then we were made to kneel with two-by-fours behind our knees. I don't know what kind of torture it is, but it does do the trick. We knelt there with everyone watching us for what seemed like ten hours. I'm sure, really it was for no more than ten minutes.[9]

However, the misery associated with protracted duration need not be outlandish, for a variety of homespun circumstances are capable of making an interval seem interminable. In fact, any form of illness or self-inflicted injury will suffice, as the following student narratives attest:

I wasn't feeling well and I went to bed around 11:30 p.m. My sleep was disturbed by fever and strange, anxiety-causing dreams. When I awakened, I felt feverish and miserable, and I felt as though I'd been in bed for . . . four or five hours. Actually, though, it had been only about an hour and a half.[10]

My mom and I got into our car to go home, which was only ten minutes away. I was a nosy seven-year-old, and I climbed into the back to see what kind of groceries we had. Slowly, I put my hand into a bag and pulled out a red thing. It looked as if it would taste good so I took a big bite. Instantly, my eyes got big, and my mouth felt as if it was on fire. I started kicking and screaming, but my mother couldn't help me. I had to sit and wait for us to get home. It seemed as if my mother was driving five miles per hour. The pain seemed to sizzle. Finally, when I got home, it felt as if thirty minutes had passed.

In addition, there are offensive experiences. Such experiences have little if any effect on us physically, but, by offending our sensibilities, they make for the kind of suffering that occasions protracted duration. For instance, I once witnessed a sociological presentation during which the researchers played a videotape from a pornographic film. Immediately after the videotape ended, the following dialogue occurred between one of the researchers and a female, middle-aged member of the audience:

"*That* was four minutes?"
"Yes, I believe about that."
"It seemed interminable. I thought it would never end."

Likewise, one of the young women we interviewed described two hours of dinner with an obnoxious date as an "eternity."[11]

Offensive experiences provide a convenient transition to consideration of the emotions that result from our bruised sensibilities. Indeed, the catalog of offensive experiences segues quickly into unpleasant combinations of disability and sentimentality. Homesickness is the easy example, as this student's account shows:

After about two months of being in England, it felt like I had been there at least a year. I was used to the culture, [and] I had a boyfriend there, [but] it seemed like I had been away from home for so long!

Norman Denzin makes the general point when he observes that one's involvement with intense emotions stretches lived duration.[12] Vehement emotions focus one's gaze on the problem at hand. This effect is portrayed with great clarity in Jack Katz's analysis of rage:

The blindness to the practical future implications of the moment gives rage, with all its fury, a soothing, negative promise that humiliation painfully lacks. This is its great comfort. Like the promise of an erotic drive, rage moves toward the experience of time suspended; it blows up the present moment so the situation becomes portentous, potentially an endless present.[13]

Still, despite the differences between rage and humiliation, the latter has the same effect on the perceived passage of time. Note, for instance, how a brief interval of embarrassment can seem like an interminable ordeal:

Halfway through the final performance and in front of a full house, Kathy simply forgot her lines. For no apparent reason, she blocked, and it was minutes before the play was on track again.

For Kathy, those agonizing minutes seemed like years.[14]

Depression is frequently cited as one of the emotions associated with protracted duration. A woman who had attempted suicide while overcome with despondency describes a "molasses-like feeling of being stuck in endless time."[15] Furthermore, as our interviews indicate, depression is often but one element in a more elaborate emotional response:

I was having trouble with my girlfriend [so] I went out to the sea-wall to think about things. I was feeling very . . . isolated and angry. Time seemed very slow. I also had a feeling of helplessness and I guess you would call it persecution.

Fearful situations provide another path to the perception that time is passing slowly. As with the other emotions, fear monopolizes one's attention; the fearful individual is mesmerized by some object of dread. For a young boy, that object could be the creature sharing his company in a dark shed:

I was locked in there . . . for maybe two minutes, but it seemed like days before my mom finally found me. While I was trapped, I discovered that I was accompanied by an extremely large spider which was residing in the corner. For the entire time, all I did was stare. I don't think I blinked even once.

And, for the principal character in Tom Clancy's novels, it could be a fear of flying:

They were only ninety minutes out of Oceana Naval Air Station at Virginia Beach. It felt like a month, and Ryan swore to himself that he'd never be afraid on a civilian airliner again.[16]

Still, the transition from suffering to emotions is not as straightforward as the discussion thus far might lead one to think. We have known for more than seventy years that keenly felt forms of pleasure are just as capable of bringing about the perception that time is passing slowly.[17] For example, intense enjoyment of music can induce protracted duration:

A certain kind of music has a strange effect on me . . . when the melody is one which begins very slowly . . . and softly. It is almost as if a single note begins somewhere in the infinite past and grows and grows in volume. I get this feeling that I am experiencing something like the infinity of time. When it happens, it's almost as if I can unconsciously see it out there all by itself . . . a single note which started long, long before I was born, maybe even before anybody was born. It's like it's been going on forever and only now catching up to me. Before I know it, I get caught up in it, and I listen and listen . . . and listen; and time stops until it's over.[18]

But when protracted duration is at issue, the pleasures of the flesh are just as potent:

I know it wasn't the first kiss I ever had, but it was my first unforgettable one. I was fifteen and I had a mad crush on a boy named Eddie. I had met him when we were both doing a play. One freezing cold day, we spent the whole day walking all over New York City, and we kissed on the Avenue of the Americas, outside a bank. When we kissed, my heart felt like it was going to burst, and the world began to swirl around me. It was the most incredible feeling, and it seemed like the kiss lasted an awfully long time.[19]

Peter Berger gives us a helpful summary: "Joy is play's intention. When this intention is actually realized, in joyful play, the time structure of the playful universe takes on a very specific quality—namely, *it becomes eternity*."[20] This statement reminds us that we are dealing with transformations in the perception of reality—transformations that, by nature or design, rivet one's attention to matters at hand. In addition, it should be obvious that suffering is not a necessary part of protracted duration, and time does not always fly "when you're having fun." Here, too, we are reminded of the inadequacy of folk wisdom.

Violence and Danger

Ordinarily, we find ourselves in situations that do not command our undivided attention. From a physical standpoint, we are always here now, but our thoughts and feelings often reflect the events of another, perhaps imaginary place and time. As a result, our attention to details of the situation at hand is typically desultory. However, when situations explode suddenly into violence and danger, the scope of one's attention narrows to immediate circumstances, with attendant effects on the perceived passage of time. These effects are consistent regardless of whether the violence is natural, accidental, intentional, or vicarious.

All too frequently, natural forces create situations of such astonishing violence that they seem to emanate from the arsenal of a vengeful god. Tornadoes, earthquakes, floods—these and other spectacular circumstances can bring about the experience of protracted duration:

Duane Grimm jumped out of his parked car and threw himself onto gravel. A tornado struck within seconds.

"The best way I can describe it is going 150 miles per hour on a motorcycle. All you feel is wind. . . ."

"The air pressure, the feeling of having your ears ready to pop, intensified to the point I thought I was going to lose my hearing," the 25-year-old said of his ordeal.

Grimm lay on his stomach about 15 feet from the car until the tornado passed 30 seconds later. He suffered a broken hip.

"I am damn lucky the tornado didn't pick me up," he said. "Thirty seconds seemed like a year."[21]

Armenians interviewed by Izvestia described in gripping simplicity the terror of the first moments of the quake.

"It was like a slow-motion movie," said Ruzanna Grigoryan, who was working at the stocking factory in Leninakan, the republic's second-largest city, when the building began to tremble. "There was a concrete panel slowly falling down."[22]

Three teens looked up to see an elderly couple banging on the windows of their car, swallowed by the swollen Mississippi. . . .

"I had climbed in the back to try to get some air, but the water was up to the ceiling," said Mrs. Crabtree, 67, who cannot swim. "It was an eternity."[23]

Subsequent to *The Wild Bunch* by Sam Peckinpah, it has become commonplace for directors to depict violence in their films through slow-motion cinematography. And, as we have seen, this cinematic cliché represents exactly the experiences of those who have survived violence in everyday life. Given the circular relationship between experience and cultural representations of experience, it should come as no surprise that, as they struggle to put extraordinary events into words, victims of violence often have recourse to the special effects of the cinema for their figures of speech.[24]

In addition, it is not uncommon for those victimized by violence to report heightened attention to self and situation throughout the episode. They may be stunned by how much information their minds can process in the moments before or during an automobile accident. For example, "Between the second I uncovered my eyes and the instant we hit the tree," one young woman we interviewed "thought of a million things." Moreover, those who survive accidental violence are frequently amazed by their detailed perception under difficult circumstances. When describing her automobile accident, another student said that "I could hear each dent and each piece of glass break." The following narrative is typical:

My first thought was, "Where did that car come from?" Then I said to myself, "Hit the brakes." . . . I saw her look at me through the open window, and turn the wheel, hand over hand, toward the right. I also [noticed] that the car was a brown Olds. I heard the screeching sound from my tires and knew . . . that we were going to hit. . . . I wondered what my parents were going to say, if they would be mad, where my boyfriend was, and most of all, would it hurt. . . . After it was over, I realized what a short time it was to think so many thoughts, but, while it was happening, there was more than enough time. It only took about ten or fifteen seconds for us to hit, but it certainly felt like ten or fifteen minutes.[25]

The characteristic experiences of intentional violence echo those of natural and accidental violence. There is a sense of extreme urgency and concentration on matters at hand. One is utterly absorbed with self and situation. Numberless thoughts and feelings flow through one's supercharged consciousness. Protracted duration is the upshot:

The whole thing didn't last more than two minutes. I'm sitting there thinking, "What am I going to do?" I'd had guns pointed at me before, but I'd never been shot at. I can't even begin to tell the things that went through my head. "Jesus, why didn't I wear my vest? Who is this crazy son of a bitch? Why is he doing this? Jesus, they don't pay me enough for this shit." A million things. Plus you're trying to count bullets.[26]

Miller knew most of Dunedin's transients. But as the 22-year-old patrolman drove away from the station, he saw one he didn't recognize, a gaunt unkempt man carrying two large bags. He pulled over and told the man he wanted to speak to him.

As he reached for his radio to report stopping a suspect, he noticed the man had dropped the bags. The man was walking toward him, arms crossed, his right hand inside his jacket.

"Let me see your hands! Let me see your hands! Let me see your hands!" Miller yelled.

The man dropped into a combat crouch.

Five years later, Miller still recalls vividly the slow motion of the next few seconds, the tunnel vision, the revolver barrel looking as big around as a grapefruit, the shots . . . one, two, three . . . hitting his gut within two fingers' width of each other. How the man pointed the gun at his head and, click, the gun misfired, leaving one live round in the chamber, and the wide, wide look in the man's eyes as the policeman he had shot arose to return fire.[27]

The perceived passage of time slows for the perpetrators as well as the victims of intentional violence. This is apparent when Jack Abbott, a recognized expert on these matters, describes the climactic experiences of a jailhouse murderer: "You go into a mechanical stupor of sorts. Things register in slow motion because all of your senses are drawn to a new height."[28] Moreover, violence can dominate our attentional resources even when we are not directly involved. A sensible concern for vigilance requires sensitivity to that which is alarming or unexpected. Thus, in some instances, the effects of vicarious violence can be attributed to a visually arresting scene. Consider, for example, the following narrative from one of our interviews:

My friend and I were riding our bikes, and we stopped to rest at a corner. As we were standing there, I noticed a car coming down the road. It was going too fast, and, as it attempted to make a turn, it flipped over. It flipped five or six times, probably taking only a few seconds, but it felt like minutes. I could see each turn clearly as it slowly approached this small tree.

In other instances, the effect of vicarious violence on the perceived passage of time appears to be derived from our capacity for sympathy and taking the role of the other:

He plays them a recording made by the Akron, Ohio, Police Department of a woman who is being raped.

After eight minutes of hearing the woman's agony, the men in Staehle's audience are silent, their eyes wide and glazed.

Some of the men ask Staehle to turn the tape off. Most can't believe it lasts only eight minutes; it seems far longer.[29]

However, lest we forget, perception is at issue—the perception of danger, the way it galvanizes the senses as well as consciousness, and its effect on the perceived passage of time. Consequently, the threat of violence is enough to bring about the experience of protracted duration. Two reporters stumble upon this discovery when they investigate an urban riot:

I saw them moving toward us and began to think I should not stop at the light. The teen-agers were smiling, as if happy to see us. One had a baseball bat and was moving toward the back of the car. Another made a signal to some friends on the other side of the street. Some of them carried rocks that looked, at least from Lynda's perspective, the size of footballs. She said later they looked as if they were moving in slow-motion.[30]

It is easy to assume that such circumstances are scary, but we must remember that fear is not a necessary part of the experiential process. When a gunman on a rampage outside the White House pauses to reload his rifle, two tourists seize the opportunity to disarm him. Later, one of them says, "'I was not afraid. Everything moved in slow motion.'"[31]

In a tangential vein, it is interesting to note that a sizeable segment of the entertainment industry is based upon our willingness to pay good money for the appearance of danger. Perhaps, like other amusements, the attraction is rooted in the liberating qualities of "reality play."[32] In any event, protracted duration often can be counted among the various experiences provoked by such "amusements." This student's account is representative:

The roller coaster . . . looked like great fun, but, as the old saying goes, looks can be deceiving. . . . The ride was to last only three seconds [but] those three seconds were my lifetime. This is the highest roller coaster in the U.S., and it is straight down. . . . As we began to fall, I relived my entire life. I saw pictures of me and my family. I saw me playing with my dog. I saw many memories that just popped up. It seems I fell forever.

And, inevitably where perception is at issue, there will be misperceived or false threats. Still, illusory danger has no less impact on the perceived passage of time. The example below comes from one of our interviews:

I was in the bathroom, and I heard someone come up the stairs. I was alone, and. . . . I had left the door unlocked because my sister would come in late. I could hear the person come up the stairs because the floor cracked. I . . . instantly focused on my watch [thinking] I could give it to the maniac so he wouldn't kill me (which was pretty stupid). It seemed like he was coming up those stairs so slowly. I thought it took about fifteen minutes [but] it only took less than a minute. I heard the maniac come toward the bathroom, and I was ready to [offer] my watch for my life. It turned out to be my brother.

So, when cataloging the sufficient causes of protracted duration, we must add violence and danger to suffering and intense emotions. Once again, we have observed marked departures from routine circumstances, heightened attention to self and situation, and efforts to translate anomalous experiences into standard temporal units. These themes will continue to emerge from later narratives.

Waiting and Boredom

The cessation of ordinary activity usually has the effect of drawing out the perceived passage of time. Consequently, temporal disorientation tends to follow in the wake of natural disasters, as displaced persons adjust to the disruption of their lives in the aftermath of destruction:

He collected what clothes were left in the ruins of his apartment . . . then he got a room at the Trade Winds Motel.

To keep from hanging around the motel, he went to his office, but nobody was working. Time stood still.[33]

William James observes that "a day full of waiting, of unsatisfied desire for change, will seem a small eternity."[34] His reference to dissatisfaction suggests captivity as the root metaphor. "Waiting casts one's life into a little dungeon of time."[35] Moreover, the fact that some people, like models, are paid handsomely for brief moments of captivity does not appear to forestall complaints:

I had the feeling they were playing a game—waiting for me to do something surprising. They'd egg me on—I'd strike various poses. Sometimes I'd feel like a dancing dog. Then Allan would shout, "Hold it!" and I'd remain motionless for what seemed like hours—I'd often have to count to a hundred and twenty before they clicked.[36]

Regardless of whether it is imposed on us by natural or social forces, waiting is viewed as an imposition; it is a sentence we serve unwillingly.

There is a relationship between waiting and boredom, but, as with suffering and the emotions, it is not a necessary relationship. We do not merely wait; we wait *for* something, and, more often than not, our emotional response to waiting is colored by what we are waiting for. There are, for example, protracted intervals during which we are tense and worried by perceived delay in the arrival or achievement of that which we desire. The following student narrative is a case in point:

I was waiting for my boyfriend for two hours. . . . At first I was relaxed, expecting people to always be a little late. Then I found myself looking at my watch more and more. People would knock on the door and I would jump up, then sit down again more agitated. At 7:00, a show came on lasting half an hour. It seemed like it lasted an hour. Every time I looked at my watch, it seemed like it had stopped, only starting again when I looked at it, as if it were playing a kind of game on me.[37]

Thus, as indicated by one of our informants, waiting can be nerve-wracking rather than tedious:

I was waiting for the professor to come in and give us the test. I had a knot in my stomach and I was concentrating on keeping myself calm so that I wouldn't draw a blank on the test. Time passed very slowly—what was actually only about five minutes seemed to be a half hour or so. I wanted time to pass quickly—just to get it over with.

Something of the same experience is reported by those who have awaited word on the fate of loved ones in peril as well as those who have, themselves, lingered in the midst of danger or distress:

Candish said they tried unsuccessfully to right the boat, and then decided to sit and wait for help. "It was definitely the longest night of my life," he said.

It was also a long night for the families of the two men.[38]

In a related vein, Tolstoy gives us a character fretting impatiently while awaiting the doctor who will deliver his wife's baby: "Three minutes passed; it seemed to Levin that more than an hour had gone by."[39]

Barry Schwartz points out that "we often judge the waiting period to be longer than it is because we then pay more attention to time than we would ordinarily do during an objectively longer active period."[40] Like the girl waiting for her boyfriend, we check our clocks (or calendars) so frequently that the passage of time is sliced into thinner and thinner increments, making it seem almost to have stopped. What is more, the actual amount of time is not at issue. An objectively long interval can seem to have passed quickly, while an objectively short interval can seem interminable. Indeed, when astronauts arrive at a transitional point on their journey to the moon, a great deal of attention to temporality can transpire within a span of minutes:

> Time seemed to slow down. Each man knew the engine must fire for the prescribed duration—no more, no less. If the engine shut down prematurely, or if it didn't deliver the proper amount of thrust, they could end up in a weird, errant orbit. If it fired even a few seconds too long, Apollo 8 would lose so much energy that it would crash into the moon. By the 2-minute mark the burn had begun to seem very long. Borman said aloud, "Jesus, four minutes?"
> "Longest four minutes I ever spent," Lovell said as the engine roared silently in the vacuum.[41]

Anticipation is a key ingredient. It follows that those who have grown accustomed to working on word-processing equipment have no trouble identifying with Tom Clancy's effort to create an air of tension: "It took twenty seconds, an eternity in computer time."[42] Clearly, the crucial factor is a violation of one's expectations. Something is taking longer than it "should," and the individual can do nothing about it except wait. In other words, waiting is a problem because one is capable of learning (and thereby recognizing departures from) the normal pace of social interaction. Waiting can be exacerbated or ameliorated by cultural arrangements, but, in a fundamental sense, it is only visible when viewed against the backdrop of temporal order and socialization.

Likewise, the boredom experienced in a given situation is relative to the normal level of interest evoked by one's activities and surroundings. Consequently, an urbane visitor to a less than cosmopolitan community may feel trapped in nearly interminable emptiness: "As far as the slow pace of St. Petersburg, she says, 'I've been here for eight weeks; it seems like eight years.'"[43] There is, then, no clean line between suffering and the tedium of suspended animation, as is well known by victims of imprisonment:

The days passed with a terrible, enervating, monotonous slowness, the tomorrows blending into weeks and the weeks blending into months. "We were about a year in Auschwitz," says Menashe, "But in Auschwitz, one day—everyday—was like 10 years."[44]

Prisoners suffer all manner of tribulations, but no memoir of incarceration fails to mention the ordeal of monotony. Of his solitary confinement, Jack Abbott says, "When you neither move nor think in your cell, you are awash in pure nothingness."[45] Malcolm Braly adds that the "hardest part of serving time is the predictability. Each day moves like every other. You *know* nothing different can happen."[46] Tim Roche concurs: "boredom remains the most serious unpleasant consequence of incarceration."[47]

Unfortunately, boredom and its effects on lived duration are not unique to the world of imprisonment. Ornithologists enjoy the various perquisites afforded by their profession, but, like other scientists, the routine tasks of their research can be quite monotonous: "The work was often dreary . . . long hours hidden in a blind, looking through spotting scopes, marking habits and movements."[48] And, if eagle-tracking has its boring aspects, then, as one of our informants reported, restaurants specializing in fast food can be depended upon for employees who experience slow time:

I was doing my . . . routine chores at McDonald's. . . . However, the crowd [left], so I didn't have to make any more food. Instead, I was told to clean up the production area. . . . The slowest part of the job would come at night, near the end of closing. Whenever I was put in charge of cleaning the utensils, time seemed to go by slowly. I was put in a room away from the other workers, where the sink was. Also, there was no clock within sight, and when I did check the time it was always much earlier than I had anticipated.

Waiting is not always boring, and boredom does not always involve waiting. Still, these concepts are related, if not identical, and the fact remains that protracted duration can be brought on by a lack of variety.[49] A character in Joseph Heller's novel, *Catch-22*, tries to exploit this fact by initiating boredom so as to make his life seem longer:

Dunbar was lying motionless on his back again with his eyes staring up at the ceiling like a doll's. He was working hard at increasing his life span. He did it by cultivating boredom.[50]

Mihaly Csikszentmihalyi provides theoretical guidance for those who, like Dunbar, would construct boring circumstances.[51] He argues that enjoyment results on those tenuous occasions when personal skills are matched by the demands of the situation at hand. Otherwise, problems arise: anxiety, when personal skills are inadequate for situated demands; boredom, when personal skills exceed situated demands. His recipe is plausible in the abstract, and it reminds us that boredom is not necessarily a response to "emptiness," but there is reason to believe that more is involved than a lack of balance between personal skills and situated demands.

Among students, classroom situations were, by far, the single most prevalent context for boredom. Thus, all too frequently, jail and class have much the same effect on the perceived passage of time. Even allowing for some poorly taught classes, however, it appears that a sizable number of students refuse to embrace what could be a challenge worthy of their intellectual skills:

I'm sitting in my math class, ignoring everything the teacher says because I don't understand the material. I feel restless, like a hyperactive child! I can't stop bouncing my foot. I keep asking, "What time is it?" to the people around me. I realize that some of my classmates are just as bored as I am. It felt like time stopped. I would think twenty minutes had gone by when actually only five had passed!

Obviously, a "boring" situation can present one with potentially engrossing circumstances. Perhaps some degree of agency figures in the phenomenon: Does one elect to be the kind of person who takes up that kind of challenge? In any event, there can be no doubt that boredom elicits the experience of protracted duration. The irony is that waiting and boredom affect the perceived passage of time in exactly the same way as violence and danger, and for exactly the same reasons.

Altered States

The term "altered states" refers to a variety of unusual forms of consciousness. Many different things fall under this rubric, but, in one way or another, all of them effect a temporary excursion from ordinary circumstances. What is more, all of them are known to bring about the experience of protracted duration. Unlike the overt drama of violence, there is often little or nothing "happening" from the standpoint of an outside observer. Unlike the seeming emptiness of waiting, altered states of awareness are characterized by a rich flow of ecstatic, albeit at times eerie or surreal sensations. And, unlike suffering, at least some altered states have pleasure as their raison d'être.

There are, for example, those who pursue chemically induced forms of transcendence by ingesting consciousness-altering substances. Consider the following passage from Thomas De Quincey's *Confessions of an English Opium-Eater*: "Sometimes I seemed to have lived for seventy or a hundred years in one night; nay, sometimes had feelings representative of a duration far beyond the limits of any human experience."[52] In a later era, his compatriot, Aldous Huxley, experienced comparable sensations during his experiments with mescaline:

And along with indifference to space there went an even more complete indifference to time.

"There seems to be plenty of it," was all I would answer, when the investigator asked me to say what I felt about time.

Plenty of it, but exactly how much was entirely irrelevant. I could, of course, have looked at my watch; but my watch, I knew, was in another universe. My actual experience had been, was still, of an indefinite duration or alternatively of a perpetual present made up of one continually changing apocalypse.[53]

Victor Gioscia provides less personal, but more systematic evidence when he summarizes his research on the effects of LSD: "[O]ur respondents indicate that . . . moment after moment is filled with delights of the most sensuous and rapturous sort, and that, for hours on end, in what seem to be vastly extended spans of time, wholly satisfying releases of ecstatic bliss are attained with magnificent ease."[54]

Like drugs, sexual ecstasy can transform social reality and, in so doing, produce a form of consciousness that is characterized by protracted duration. Murray Davis suggests that the transformational effects of sexual ecstasy are derived from the way it narrows the stream of consciousness: "Those who leave everyday reality to enter erotic reality . . . become less attentive to both spatial (distant) and temporal (past and future) extremities but more attentive to their centers (local and present)."[55] One of his informants observes how this affects the perceived passage of time: "The best moments in sex come when both lovers really seem to merge into one. You know, those moments that seem to go on forever. . . ."[56]

In another case of art imitating life, the directors of films typically see to it that audiences can differentiate dreams and other fantasy sequences from the wide-awake experiences of their characters. For example, in his review of a film, one critic notes that "during the fantasy sequences . . . Jim's toy plane seems to stay aloft forever. . . ."[57] This is as much a cinematic convention as the handling of violence, and for exactly the same reason—it mirrors real human experiences. Sturt points out "that long dreams, appearing to last some hours, may occur in a few minutes or seconds."[58] Similarly, one of our informants enjoyed clandestine mental vacations from the rigors of class: "The daydreams seemed to last a half hour, but were no more than five minutes long." Another student gave us this equally incriminating report on inappropriate attention to matters at hand:

My boyfriend and I began a conversation right after we had been hugging. I realized in the middle of the conversation that, for what seemed like a very long time, I had been studying an image of a colorful room. The conversation had lasted for about two minutes, but it felt like I had been studying the image in my mind for about fifteen minutes.

Thus, dreams and other forms of fantasy allow one brief excursions to a wonderland of alternative worlds and protracted duration.

But are all dreams and dream-like experiences "mere fantasies"? There are, of course, people who believe that at least some of these episodes involve the actual experience of a supernatural or alien reality. The validity of their claims does not concern us, but their experiences are relevant because they are often marked by the perception that the passage of time has slowed drastically or even stopped altogether. Consider, for instance, the following description of a close encounter with an alien presence:

One afternoon in the late seventies, while she was caring for her three oldest children, Parnell looked outside and saw a UFO. She blushed as she described it to me: "It was bright, yellowish-white, and shaped like a flattened football." Parnell didn't notice how big or how far away it was, but she felt "light filling my head and time being eternal."[59]

Curiously, apart from the UFO itself, her story resembles those of persons who have been diagnosed as clinically dead, but were subsequently revived to tell the tale. Recounting her near-death experience, Betty Malz says that "there seemed to be no passing of time."[60] And when a reporter interviewed Kenneth Ring, a student of near-death experiences, he summarized his findings as follows:

It is an experience that has been reported by thousands of people who either have come very close to physical death or have actually passed into a temporary state of clinical death.

These people tell of a common pattern of experience: A sense of extreme peace and well-being, a sense of being separated from the physical body.

They speak of being able to gaze down at their physical body as if they were spectators. There's often a sense of moving through a dark space—sometimes described as a tunnel—and encountering a beautiful, warm, brilliant light that seems to engulf and surround the individual. One individual gave this description to me: "It was eternity. It was like I was always there and would always be there, and that my life on earth was just a brief incident."[61]

In a general sense, these stories are evocative of a wide variety of ecstatic episodes associated with spiritual or religious revelation. As Tom Robbins puts it, "In mystic illumination, as at the speed of light, time ceases to exist."[62]

By degrees, we have moved from the voluntary and pleasurable escapades of drug-taking literati to the involuntary but not quite unpleasant rapture of those who take uncanny excursions of one kind or another. Thus, altered states are not all fun and games; there are serious, indeed, profoundly frightening experiences in this category. Some of them are stupor-like states of semiconsciousness brought about by various forms of stress, exertion, or deprivation. Two of our student informants provide the following accounts:

During the church service, I passed out and felt like I was dreaming. Even though my eyes were open, I couldn't see; all I dreamed about was passing out in church. It seemed like I was asleep for twenty minutes, but when I came to, it had only been around thirty seconds. When I woke up, there was a crowd around me and I was very embarrassed.

I had just come into the weight room from running a couple of miles and the other players and I were instructed to lift weights for a little while. I had just begun when I started to feel sick, so I sat down on one of the benches in the weight room. I began to feel worse and slowly the scenery around me started to fade out. I got up and started to walk, and everything went completely black. I kept walking from memory of the layout of our locker room. I remember one person trying to talk to me and his voice was very slow and dragged [out] to where I couldn't understand it. I felt my way to a bench and lay down for a while. Slowly my vision came back and I felt fine. The guys in the weight room said I had been gone for only about five minutes, but it seemed like an eternity.

The less ephemeral genres of "mental" disorganization can be even more troubling. And again, there is distortion of the

perceived passage of time in the form of protracted duration. Witness, for example, Eldridge Cleaver's description
of his "nervous breakdown":

For several days I ranted and raved against the white race, against
white women in particular, against white America in general.
When I came to myself, I was locked in a padded cell with not
even the vaguest memory of how I got there. All I could recall was
an eternity of pacing back and forth in the cell, preaching to the
unhearing walls.[63]

In related fashion, Morag Coate gives us a first-person account of a psychotic interlude:

I went back into my own room and got into bed, but now I could
not sleep, and this was dangerous for the room was filled with an
unearthly light and my hand cast no shadow on the wall. The
church spire which was a landmark from my window had disap-
peared, and time was passing at an altered speed. Time was
stretched out like an elastic band, each minute of it was at once
thinner and longer than usual. At last the stage was reached when
external time ceased altogether and only I lived on. I could prove
that by the fact that my watch stopped.[64]

We have examined many unusual forms of consciousness in this section of the chapter. Whatever differences
there may be among them, these forms of consciousness
can be conceived of as more or less temporary excursions
from ordinary circumstances. They are occasioned by
drugs, sex, dreams and other kinds of fantasy, UFO sightings, near-death episodes, spiritual or religious ecstasy,
fainting and other kinds of stupor, nervous breakdowns,

and psychotic interludes. Each in its own way is capable of bringing about the perception that time is passing very slowly.

Concentration and Meditation

Thus far, it may appear that emotions of one kind or another are implicated in those circumstances which make for the experience of protracted duration. Nevertheless, as we are about to see, it is not emotions themselves, but the way they facilitate concentration—a narrowing of focus—that is crucial to the effect they have on the perceived passage of time. In a sense, we have been addressing concentration all along. Now, however, we take it up in, one might say, its "pure form."

Put differently, one need not be overwhelmed by emotion in order to perceive that time is passing slowly. On the contrary, the experience of protracted duration can be induced by deep cognitive and perceptual immersion in one's conduct or circumstances. Borrowing Erving Goffman's nice phrase, this level of concentration can lead to "an honest unawareness of matters other than" those which are pertinent to one's own activity or situation.[65] Such episodes are prevalent in sports where emotional self-indulgence is detrimental to performing at peak capacity:

At times, and with increasing frequency now, I experience a kind of clarity that I've never seen adequately described in a football story. Sometimes, for example, time seems to slow way down, in an uncanny way, as if everyone were moving in slow motion. It seems as if I have all the time in the world to watch the receivers run their patterns, and yet I know the defensive line is coming at me just as fast as ever.[66]

It is not the controlled violence of football that produces the temporal effects. In the following passage, a female race car driver gives us a glimpse of her stream of consciousness:

Floor the throttle. Shift up . . . shift down. Shift 30 to 40 times a lap—100 laps, say 3500 shifts. . . .

Really rolling now. Each lap, a few cars sliding backward—the right direction. The experience of the last five years plus the shared knowledge of the guys who have been so helpful. All of it jammed into this overloaded time frame. Elastic time. Seconds become hours, and minutes an eternity.[67]

What is more, we can assume that she is weaving in and out of traffic at an extraordinary rate of speed, all the while avoiding mishaps, watching her instruments, and plotting strategy. Here again, the irony is that time is made to pass slowly by arrangements which are meant to make something happen quickly.

Unexpectedly, athletes can find themselves in an almost eerie zone of concentration. This "zone" is characterized by an absence of emotion, greatly enhanced powers of perception, and intuitive knowledge of what needs to be done:

Today, thanks to an incendiary fourth-quarter performance by Reggie Miller and a shocking disintegration by the New York Knicks, the Pacers are within one victory of their first NBA final.

Miller, with 25 of his 39 points in the final period, including an NBA-record 5 of his 6 three-pointers, almost single-handedly led the Pacers to a 93–86 victory Wednesday night over the Knicks.

"Everything felt like it was in slow motion," said Miller, who made 14 of 26. "You see plays before they happen. You read defenses as soon as the ball's coming your way. You know what your defender's going to do before he does it."[68]

Picabo Street, an Olympic skier, describes a similar form of anticipation:

"You know, on a run, after you pass a certain speed, it all slows down," she explains. "Your mind is like two sections down the course. You're doing one thing and thinking what you have to do ahead."[69]

Presumably, there is an emotionally tinged horizon of concern for the quality of one's performance, but this provides, at most, a distant backdrop for one's efforts. In the immediacy of a seemingly "perpetual present," one's consciousness is dominated by concentration on the pressing matters at hand. Indeed, from a subjective standpoint, the flow of time can almost be halted by meticulous attention to detail: "It is over in a flash: a heart-stopping .82 sec. 'But when you're up there, it feels like forever,' says U.S. Gymnast Mitch Gaylord of his specialty, the Gaylord II."[70]

However, we need not restrict ourselves to sports. Other kinds of activity require greatly elevated levels of concentration, and they generate comparable distortion in the perceived passage of time. Consider the task of driving under extremely challenging conditions. Joan Didion claims that the Santa Monica Freeway provides an opportunity to partake in an extraordinary form of attention to one's immediate circumstances:

Mere driving on the freeway is in no way the same as participating in it. Anyone can "drive" on the freeway, and many people with no vocation for it do, hesitating here and resisting there, losing the rhythm of the lane change, thinking about where they came from and where they are going. Actual participation requires a total surrender, a concentration so intense as to seem a kind of narcosis, a rapture-of-the-freeway. The mind goes clean. The rhythm takes over. A distortion of time occurs, the same distortion that characterizes the instant before an accident. It takes only a few seconds to get off the Santa Monica Freeway at National-Overland, which is a difficult exit requiring the driver to cross two new lanes of traffic streamed in from the San Diego Freeway, but those few seconds always seem to me the longest part of the trip.[71]

In addition, there are occupations that involve ever-problematic circumstances, and yet the slightest mistake can result in disaster. For example, complete absorption in an effort toward precision is critical to those whose work entails parachuting into the sea at night on rescue missions:

"The jump lasted 30 seconds," Westmoreland said. "It seemed like an eternity. You take a lot in that 30 seconds. If you're not right next to your partner, you could lose him in a second flat.[72]

In fact, it is possible to rise above an altogether reasonable concern for one's own illness when duty calls and careful attention to detail is necessary. One of our informants reported feeling sick the night she was to dance for a fellow student's senior lighting thesis. Nonetheless, her "concentration took over." The dance "was about three minutes in length, but it felt like it took hours to complete." And yet, one can also be caught up in or engrossed by one's circumstances for reasons that are utterly self-serving. There is, for instance, a scene in Dostoevsky's *Crime and Punishment* where a character feigns nonchalance while under surveillance:

Someone was standing stealthily close to the lock and just as he was doing on the outside was secretly listening within, and seemed to have her ear to the door. . . . He moved a little on purpose and muttered something aloud that he might not have the appearance of hiding, then rang a third time, not quietly, soberly and without impatience. Recalling it afterwards, that moment stood out in his mind vividly, distinctly, forever; he could not make out how he had such cunning, for his mind was as it were clouded at moments and was almost unconscious of his body. . . . An instant later he heard the latch unfastened.[73]

Like surveillance, certain situations seem to have a natural affinity with concentration. This is especially so when, like this student, we are called upon to cooperate stoically in uncustomary practices:

I was in the clinic to get many vials of blood taken out. I sat in a chair, and the nurse asked me to roll up my sleeve. She tightened a tourniquet around my arm. Time was slowing. Then she found a plump vein and injected it with the needle of a syringe. Slowly my blood filled the vial attached to the end of the syringe. As it filled, she replaced it with another. Again, I watched as my blood rose in the vial. I watched her fill four vials, and every one seemed to take forever.

As with our previous narrators, it is important to note how this young man displays emotional detachment in tandem with a narrowly focused gaze. He is fascinated, but not otherwise aroused.

Clearly, circumstances can arise which so monopolize cognitive resources that one's attention is restricted to matters at hand. However, we should not leave off discussion of this topic without first pausing to consider willful acts of concentration; that is, efforts to focus consciousness which reflect one's intentions rather than the exigencies of the immediate situation. We cannot ignore the role played by intentionality, for, as Virginia Woolf shows us, the simple act of centering one's cognitive resources can have profound effects on the perceived passage of time:

[W]hen a man has reached the age of thirty, as Orlando now had, time when he is thinking becomes inordinantly long; time when he is doing becomes inordinantly short. Thus Orlando gave his orders and did the business of his vast estates in a flash; but directly he was alone on the mound under the oak tree, the seconds began to round and fill until it seemed as if they would never fall. They filled themselves, moreover, with the strangest variety of objects. For not only did he find himself confronted by problems which have puzzled the wisest of men, such as What is love? What friendship? What truth? but directly he came to think about them, his whole past, which seemed to him of extreme length and variety, rushed into the falling second, swelled it a dozen times its natural size, coloured it all the tints of the rainbow and filled it with all the odds and ends of the universe.[74]

In short, human beings can bring their thoughts to bear on matters which do not concern the world within reach.

Meditation is an extreme version of concentration, and, as one would expect, there is evidence indicating that it has an extreme effect on the perceived passage of time. As Edward Hall writes, "The slowing down of brain waves and the heart and respiratory rate during meditation have produced instances where people reported that 'time stood still.'"[75]

Shock and Novelty

Shock and novelty produce effects that resemble those of concentration. There is consolidation of mental resources into exclusive regard for the source of surprise or bewilderment. And again, the resulting devotion to detail prolongs the experience of lived duration. Moreover, like Charles Darwin, those who find themselves in shocking circumstances are often impressed by the enormous amount of information the mind can process within a relatively brief interval of time:

[O]nce, whilst returning to school on the summit of the old fortifications round Shrewsbury, which had been converted into a public foot-path with no parapet on one side, I walked off and fell to the ground, but the height was only seven or eight feet. Nevertheless the number of thoughts which passed through my mind during this very short, but sudden and wholly unexpected fall, was astonishing.[76]

The experience of falling unexpectedly is instructive, because its effects on lived duration seem to occur in advance of any particular emotional response. One of our informants told us about trying to stand on a ball. The ensuing fall "felt like it took about thirty seconds, but actually . . . took about two seconds." What is more, during the fall, he "felt lightheaded but safe, as if someone were laying me down to sleep." Under similar conditions, another student found his own lack of emotional response remarkable:

We played around as a group of friends normally do. I began to climb a tall tree to impress the others. The tree had small, weak branches, and, as I began to climb, a branch broke and I lost my footing and fell. The fall was probably only about five to six seconds, but it actually felt like two or three minutes. During the fall, I felt free but not scared of being hurt (which I thought would be the normal reaction).

There is, then, general arousal or excitation, a sometimes distant horizon of concern for one's well-being, but no particular emotional response to shocking circumstances.

Arlie Hochschild suggests that emotions function as something like a signaling system by helping the individual to construe the "self-relevance" of events.[77] The data on shock and novelty provide corroboration for her perspective. Frequently, lived duration is perceived to slow in that seemingly prolonged moment during which the individual tries to assess the meaning of shocking circumstances and arrive at an appropriate response. When, for example, Malcolm Jones finds that his bicycle has been stolen, surprise and interpretation precede his emotional reaction: "At this point, in the elongated second of time that it took me to take in the scene, I suppose I was supposed to get angry."[78] In addition, there are times when one does not know what to make of one's circumstances. Consequently, no particular emotion arises because one is uncertain about the self-relevance of events. Still, the individual finds the circumstances shocking, which results in the experience of protracted duration:

After the bodies are dragged out, without any command to leave the perimeter, the troopers start wandering in with their rucksacks. Out of nowhere, all these Instamatic cameras began to appear and flashbulbs began to pop simultaneously. I had a very strange feeling as if I was projected somewhere outside of it. Pop. Pop. Pop. I saw it as if it were in pantomime, slow motion. All these guys reaching gracefully and deftly into some hidden pocket in their fatigues, the strobe light effect of the flashcubes. They're smiling these big smiles of great joy, like something wonderful had just happened.[79]

Shocking circumstances can be good or bad, right or wrong. Surprise, not morality, is at issue, and one may not know what to feel. The key factor is a sudden encounter with the unexpected; such conditions have an utterly systematic effect on the perceived passage of time:

Dennis Franz was pretty confused Monday.

Confused why his series *NYPD Blue* didn't win an Emmy as best drama and equally confused why he won as best actor. . . .

"Surprised is an understatement. I was overwhelmed," he said. "It took Meredith Baxter about an hour to say my name. When she did there, time sort of stopped and there was a big thud in my chest. I wanted to wrap my arms around everybody in that room and give them a big kiss."[80]

Likewise, with the onset of interracial sexuality, the young woman in the following narrative finds herself in alien circumstances:

Time seemed normal when we were dancing. We went outside for fresh air, and time slowed down a little. I guess that is from getting away from the fast music. The other person started kissing me, and that's when time really slowed down. Then I said, "Stop, wait, this can't happen," and time really seemed to kind of stop. I guess the situation was against what is normal for me.

And objects of one kind or another can be so novel, so unusual, that their mere presence holds in thrall nearly all of one's attentional resources:

Stewart, who has never driven the car, watched with a grin as Randy Styf, a full-time carpenter and part-time auto restoration expert, drove the car out of the garage—a process that seemed to take a week as the gleaming scarlet hood over the straight-8 engine kept coming for an almost cartoonishly long time.[81]

This capacity to become engrossed in or carried away by one's perception of the unexpected may help us to account for the related experience of protracted duration: "When witnessing a real boomerang flying, one is often left with the impression that the flight lasted about half a minute, whereas its real duration was a mere eight seconds or so."[82]

Sometimes we come upon shocking or novel circumstances quite by accident. Moreover, for any number of reasons, these circumstances may be undesirable. Not infrequently, however, we seek out shocking and novel circumstances for the piquancy they can add to life.[83] Indeed, many leisure activities involve a diligent hunt for that moment of adventure which is marked by protracted duration:

The end of the rod lazily bounced with the soft chop of the Gulf. Suddenly, it snapped down almost double and froze. The drag on the reel howled as line was pulled out. Without warning, a huge, bluish-green fish with a short spear thundered into the air like an underseas missile. After what seemed like an eternity of headshaking and tailwalking, it crashed back into the water.[84]

Perhaps leisure activities are meant to recapture, if only for a while, the temporal experience of youth. Edward Hall notes that at "the age of four or six, a year seems interminable; at sixty, the years begin to blend and are frequently hard to separate from each other because they move so fast!"[85] Like Hall, David Maines attributes this tendency to the mathematics of the life course: "[T]he duration of a period of time is relative to the total duration of one's life. Thus, 1 year out of 20 is correspondingly longer than 1 year out of 70."[86] However, there is another possibility. It may be that, unlike their jaded elders, time seems to pass slowly for young people simply because so much of their experience is new, striking, and memorable; in short, there is more novelty in their lives, with its attendant effects on the perceived passage of time. In his "Excursus on the Sense of Time," Thomas Mann alludes to this latter interpretation:

We are aware that the intercalation of periods of change and novelty is the only means by which we can refresh our sense of time, strengthen, retard, and rejuvenate it, and therewith renew our perception of life itself. Such is the purpose of our changes of air and scene, of all our sojourns at cures and bathing resorts; it is the secret of the healing power of change and incident. Our first days in a new place, time has a youthful, that is to say, a broad and sweeping, flow, persisting for some six to eight days. Then, as one "gets used to the place," a gradual shrinkage makes itself felt. He who clings or, better expressed, wishes to cling to life, will shudder to see how the days grow light and lighter, how they scurry by like dead leaves, until the last week, of some four, perhaps, is uncannily fugitive and fleet.[87]

Whatever the case may be, it is indisputable that shock and novelty bring about the experience of protracted duration. Any encounter with the unexpected provokes greater subjective involvement with one's immediate circumstances. So, regardless of whether it emanates from distasteful shock or piquant novelty, the upshot is an unmistakable slackening in the pace of lived time. And, of course, there is some irony in the fact that the mind's capacity to work quickly is fundamental to the perception that time is passing slowly.

The Distribution of Sufficient Causes

Chapter 2 traced the outline of variation in the perceived passage of time. This chapter represents an effort to fill in some of that outline. We have examined a great many situations that differ from each other in a variety of idiosyncratic ways. Yet, despite their contextual differences, these situations all share at least one thing in common: they suffice to bring about the experience of protracted duration—that is, the perception that time is passing slowly. Thus, the empirical foundation for this chapter is provided by 705 first-person accounts of situations in which the passage of time was perceived to slow. What is more, although each one of these narratives is, in some ultimate sense, unique, they can be classified into six broad themes. It seems reasonable to assume that these themes correspond to a set of underlying factors. We can conceive of these underlying factors as the sufficient causes for protracted duration. By way of concluding this chapter, let us briefly consider the distribution of these sufficient causes.

As I noted earlier, two methods were used in obtaining the 705 narratives. The first of these involved locating 389 relevant stories in the newsmedia and books of the autobiographical genre. For the sake of convenience, we can think of these data as having been "found" in previously published sources. The second of these methods involved conducting interviews with 316 students enrolled in several sections of an introductory sociology course. Table 1 presents the distribution of sufficient causes in each of these subsets as well as the combined distribution.

There is nearly complete unanimity among the three distributions. Looking first at the combined distribution, one can see that, in descending order of frequency, the six themes are (1) suffering and intense emotions, (2) violence and danger, (3) waiting and boredom, (4) altered states, (5) concentration and meditation, and (6) shock and novelty.[88] With regard to the rank order of categories, the interviews distribution is a mirror image of the combined distribution. However, the found distribution differs from the other two in that violence and danger is the largest category, followed by waiting and boredom. Respectively, these differences may reflect nothing more than the newsmedia's appetite for

TABLE I

Distribution of Sufficient Causes by Data Set

	Found	Interviews	Combined
Suffering and Intense Emotions	24.2	34.2	28.6
Violence and Danger	28.5	25.9	27.4
Waiting and Boredom	25.2	20.6	23.1
Altered States	10.3	7.6	9.1
Concentration and Meditation	5.1	5.7	5.4
Shock and Novelty	4.6	4.1	4.4
Other	2.1	1.9	2.0
Total	100.0	100.0	100.0
	(n=389)	(n=316)	(n=705)

excitement and the ease with which one finds memoirs of imprisonment. In any event, the similarities among these three distributions greatly outweigh their differences. The top three categories are the same in each of the distributions, and they account for more than 75 percent of the data. The bottom three categories account for 20 percent or less of each distribution.

Nonetheless, as we turn our attention to theory construction, the smaller categories are just as important as the larger ones. Thorough familiarity with the full range of variation creates an empirical gauntlet through which only the hardiest of hypotheses may pass. Theory construction is directed toward comprehensive explanation and generalization beyond the data at hand. It follows that theories are evaluated in terms of their empirical reach. With its extensive survey of empirical materials, this chapter sets the stage for the formulation of a theory that accounts for the full range of variation in the perceived passage of time.

[4]

Theory Construction

IN THE PRECEDING chapter, we familiarized ourselves with the circumstances that bring about the experience of protracted duration. Having done so, we are now in a position to assess the validity of Schutz and Luckmann's statement that temporal consciousness "alters with transitions from one province of reality . . . to another, as well as . . . with transitions from one situation to another within the everyday life-world."[1] There is unequivocal support for their statement. Over and over again, we have witnessed testimony to the effect that the perceived passage of time slows within the context of unusual circumstances. Moreover, we have seen that this modification in temporal consciousness is invariably provoked by the displacement of one domain of social reality by another or by conspicuous transitions in the definition of the situation. Without exception, the sufficient causes of protracted duration implicate recognized departure from ordinary circumstances.

Still, the substantiation of Schutz and Luckmann's statement merely begs the question that prompts the present investigation: *How* does transition from one realm of social reality to another condition the perceived passage of time? This question changes the emphasis from description to explanation. As a provisional response to this line of inquiry, we must try to abstract from the foregoing narratives the underlying features they share in common. We can think of these features as components of the essential process that makes for the experience of protracted duration. In so doing, we should bear in mind that the distillation of an essential or necessary cause can never be anything more than a tentative and inductive extrapolation beyond the data at hand. Put differently, the resulting construct will be a working hypothesis and, as such, little more than a springboard for further research.

Before proceeding to their common features, however, it may be useful to review the narratives with an eye toward dispensing with those working hypotheses that do not fit the empirical materials. To begin with, it is apparent that the presence or absence of particular emotions is of no general consequence. Some of the narrators expressed sheer delight, while others spoke of unmitigated horror. It should be obvious, then, that the pleasantness or unpleasantness of the incident is irrelevant. Nor does the liveliness of the setting play any systematic part in the outcome. This is evident in the fact that certain vignettes were crowded with simultaneous activity, while others were empty stretches of monotony. Likewise, no significance can be attributed to the degree of volition, because some of these situations were actively courted, while others were forcefully imposed. These are all analytical dead ends.

This chapter represents an effort to construct a theoretical model that encompasses the full range of variation in the perceived passage of time. To that end, the conceptual model must resolve each of the three paradoxical aspects of lived duration: time is perceived to pass slowly in situations with abnormally high or abnormally low levels of overt activity; the same interval of time which is experienced as passing slowly in the present can be remembered as having passed quickly in retrospect; and some busy intervals are experienced as passing slowly, while others are experienced as having passed quickly. In addition, the conceptual model must explain the conditions that give rise to each of the three elementary forms of lived duration: the perception that time is passing slowly (i.e., protracted duration), the perception that time is passing in step with clocks and calendars (i.e., synchronicity), and the perception that time has passed quickly (i.e., temporal compression). The balance of this chapter is directed at construction of that theoretical model.

Protracted Duration

Two concepts were especially helpful in my efforts to formalize a theory of lived time. J. David Lewis and Andrew Weigert use their concept "time embeddedness" to describe "the fact that all social acts are temporally fitted inside of larger social acts."[2] For instance, when we meet a colleague in the hall during finals week, we know that time available for chatting is limited by the deadline for grades. In related fashion, H. Wayne Hogan defines "stimulus complexity" as the totality of "the individual's environmental experiences."[3] It follows that for the person in a given setting, the higher the degree of time embeddedness, the higher the degree of stimulus complexity. Both concepts refer to objective features of the situation, but they also have implications for the individual's assessment of his or her place in that situation.

From an objective standpoint, situations can be arrayed along a continuum between those characterized by extremely high levels of stimulus complexity and those characterized by extremely low levels of stimulus complexity. Another way of envisioning this distribution is to differentiate situations which, objectively speaking, are eventful from situations which an outside observer would classify as uneventful. Working independently of each other, two of my research assistants and I attempted to classify the narratives into three categories along this continuum: low, moderate, and high stimulus complexity. A specific narrative was to be classified as having a moderate level of stimulus complexity if it displayed roughly the same eventfulness as ordinary interaction in everyday life. Narratives displaying greater than normal and less than normal eventfulness were to be classified as having high and low levels of stimulus complexity respectively.[4]

Although it was our intention to classify the narratives into three categories, a bifurcation quickly became apparent as we examined the 705 texts: without exception, the texts were accumulating in the two extreme categories. We found that 42.8 percent of the narratives described situations of low stimulus complexity, while 57.2 percent described situations of high stimulus complexity. Examples of situations classified as having low stimulus complexity included being put "on hold" after telephoning someone, time in a sensory deprivation tank, waiting for one's plane to land, a day when few customers enter the store, and solitary confinement while imprisoned. Examples of situations classified as having high stimulus complexity included earthquakes and tornadoes, combat and other forms of interpersonal violence, the use of hallucinogenic drugs, automobile accidents, and guarding the ball handler in the final seconds of a close basketball game.

Thus, working backward from the experience of protracted duration to those occasions which provoke that experience, we find that the latter consist of relatively eventful *or* relatively uneventful situations. Our data contain a somewhat larger number of eventful than of uneventful situations, but this is probably an artifact of recall in that vivid memories favor eventful episodes. In any event, the fact remains that protracted duration is not associated with situations characterized by the normal eventfulness of social interaction in everyday life. And why should we expect that to be the case? Protracted duration is a temporal anomaly; as such, we should not be surprised to discover that it is experienced exclusively in unusual circumstances. In short, the first paradox of lived duration is that time is perceived to pass slowly in situations with abnormally high *or* abnormally low levels of overt activity. Either way, the individual finds the situation problematic. How, then, is the flow of attentional resources altered by transition to problematic circumstances?

As you read George Herbert Mead's lectures in *Mind, Self, and Society*, you repeatedly encounter people who have problems. There is the person who "has a nail to drive . . . reaches for the hammer and finds it gone."[5] There is the person who "cannot get a lock to work."[6] And there is the "man walking across country [who] comes upon a chasm which he cannot jump."[7] Mead's imagery reflects one of the cardinal tenets of his theoretical framework: "Reflective thinking arises in testing the means which are presented for carrying out some hypothetical way of continuing an action which has been checked."[8]

John Dewey's position strongly resembles that of Mead: "[I]t is a commonplace that the more suavely efficient a habit the more unconsciously it operates. Only a hitch in its workings occasions emotion and provokes thought."[9] There is, however, an obvious difference in their respective positions. Dewey argues that, in addition to thought, emotion is also galvanized by problematic circumstances. For Mead, role taking is crucial to the emergence of self-consciousness, and this is, of necessity, a cognitive process because emotion "does not directly call out in us the response it calls out in the other."[10]

The writings of Mead and Dewey suggest that the volume of what William James calls the stream of consciousness varies in response to the relationship between self and situation.[11] Mead and Dewey agree that, when confronted by problematic circumstances, increased attentional resources are brought to bear on whatever is blocking otherwise routine and relatively unself-conscious conduct. In a novel or difficult situation, there is heightened self-consciousness of tacit social action in the form of thinking and feeling. Cognition and emotion are different facets of subjective experience, but both cognition and emotion are amplified to the degree that habit and custom are challenged by problematic circumstances. In any event, the issue at hand concerns the total volume of experience and not analysis of its variegated content. Given the relationship between self and situation, it seems reasonable to conclude that variation in the perceived passage of time is conditioned by variation in the volume of experience.

The terms "self" and "situation" refer to subjective and objective factors, and both are implicated when we ask how unusual circumstances change the flow of attentional resources. Samuel Johnson provides us with an important clue when he assays the effect of one very problematic situation: "Depend upon it, Sir, when a man knows he is to be hanged in a fortnight, it concentrates his mind wonderfully."[12] Social psychologists have generalized Johnson's observation into a central tenet of symbolic interactionism: "*Self-consciousness* arises when habit and custom cannot guide behavior."[13] Furthermore, Erving Goffman points out that, given unusual circumstances, increased attentional resources are also brought to bear on one's environment: "In the face of ambiguities or incongruities, the puzzled or suspicious individual . . . will sharply orient to his surround and maintain vigilance until matters become clear."[14] Therefore, problematic circumstances provoke intensification of those attentional resources directed toward the internal experience of self (consciousness) and the external experience of situation (perception).

These considerations bring us back to the first paradox of lived time: protracted duration is experienced in situations characterized by extremely high or extremely low levels of stimulus complexity. Analysis is called for which resolves the paradox by explicating this bifurcation in the data. And, by the same token, this analysis would identify the conditions that account for the perception that time is passing slowly. Through inductive theory construction, it is possible to formulate a conceptual model that encompasses the forked configuration in our empirical materials. The formalization of this theoretical model begins with specification of five sequential factors that produce the experience of protracted duration. These factors are common to all of the narratives that constitute our data.

First, there is a context that is characterized by *extreme circumstances*. The situations that engender protracted duration fall into one of two categories: those with unusually high levels of overt activity, and those with unusually low levels of overt activity. These situations are not as different as they might seem to the casual observer. Indeed, they share an underlying unity that results in their common effect on the perceived passage of time. Solitary confinement and hand-to-hand combat share the fact of extremity. That is, without exception, the empirical materials depict fairly severe departures from the more habitual realities of everyday life. The anomalous experience of protracted duration casts doubt upon what Schutz and Luckmann call the "natural attitude" of the normal, wide-awake adult who typically lives in a taken-for-granted universe of common sense.[15] Their use of the phrase "common sense" refers to the world that is within common reach of our senses. This shared version of the world serves as the foundation for the establishment and maintenance of intersubjectivity. It is, among other things, the basis for collective efforts to standardize temporality.

Second, when people are confronted with extreme circumstances, they experience an increased *emotional concern* for understanding the nature of their situation. One finds oneself positioned in a setting that has become problematic. Uncanny occasions foster anxiety in human beings because such circumstances erode their taken-for-granted "trust" in familiar routines and interpretations.[16] Consequently, people feel a strong need to account for existing conditions. Transformations in the definition of the situation threaten to undermine mutuality by abandoning us in private realms of incommensurate impressions. They seize our attention because definitions of the situation become at least somewhat problematic during such transitions. Schutz and Luckmann have described the emotional response to unfamiliar circumstances as a "shock" that accompanies the transition from one realm of being to another:

The province of meaning of this world retains the accent of reality as long as our practical experiences confirm its unity and harmony. It appears to us as "natural" reality, and we are not prepared to give up the attitude that is based upon it unless a special shock experience breaks through the meaning-structure of everyday reality and induces us to transfer the accent of reality to another province of meaning.[17]

As we have seen, however, the affective response to strange or challenging circumstances does not take the form of any particular emotion. Rather, there seems to be a general state of arousal or what we can refer to as a horizon of concern. In other words, the individual cares deeply about the situation and his or her place in it.

Third, the shock of transition to extreme circumstances heightens *cognitive involvement* with self and situation. Examination of the empirical materials reveals ubiquitous amplification of perception and self-consciousness on the part of those who find themselves mired in protracted duration. They betray extraordinary concentration on the details of their immediate circumstances as well as the dynamics of their own reflexive thoughts. Galvanized by emotional concern for the implications of the situation, individuals become caught up in a special subjective "effort" at assembling the information that will enable them to make sense of unusual conditions. As a result, the individual becomes sensitive to a larger than customary proportion of the infinite information that is available in any setting. One's cognitive involvement with definition of the situation varies according to the nature of the occasion, but, as Goffman notes, concentration on the establishment of meaning "is deep only when there is sudden trouble to avoid."[18] Thus, in problematic circumstances, one's cognitive involvement with self and setting undergoes dramatic efflorescence. As involvement increases, so does attention to one's situation;

and, as attention increases, so does the density of experience per standard unit of temporality (e.g., seconds, minutes, hours, and so forth). If, with Mead, we conceive of subjective processes as tacit action, then cognitive involvement intensifies a form of embedded activity.[19]

Fourth, heightened perspicacity toward one's own subjectivity and surroundings generates *stimulus complexity* even if the setting is not characterized by a wealth of overt activity. Stimulus complexity figures in the experience of lived time, but it is not given directly in the objective qualities of the situation. On the contrary, embedded activity is conditioned by the extent of emotional and cognitive involvement among those individuals who enact the episode. Every situation is composed of much more than any one person could ever hope to acknowledge.[20] The extent of one's involvement with the proceedings governs whether one attends to more or less of the perceptual possibilities. In turn, this degree of attention determines the actual degree of stimulus complexity that is experienced by the individual. Put differently, increased perception and information processing make for greater embedded activity, albeit of a tacit or subjective form. Lived time slows when human beings become enthralled by that which they would typically ignore. This is indicative of the fact that their "relevance systems" have been drastically enlarged in response to the pressing need to comprehend the transition to extreme circumstances.[21] What was not relevant now is relevant, and, ontologically speaking, what was not, now is. From this perspective, the known world is contingent on matters at hand.

Fifth, increased stimulus complexity fills standard units of temporality with a *density of experience* that far surpasses their normal volume of sensations. According to the clock, every minute is the same, but the perceived passage of time is characterized by variation in response to one's involvement with immediate circumstances. Like the gondola cars of freight trains, standard units of temporality are identical to one another in structure but open on top so that they can be filled with an unspecified amount of ore. So too, a minute can be "loaded" with an enormous quantity of experience or very little, but, whatever the volume, it is not simply determined by the objective activity that transpires during a particular interval. Where human experience is at issue, standard units of temporality that seem empty to an outside observer may actually carry a great deal of subjective activity on the part of those who are concerned with their circumstances. For example, from an objective standpoint, little is happening when we are forced to wait, but, by paying more attention to time itself than is normally the case, we burden standard units of temporality with a mass of subjectivity that makes the interval seem much longer than it really is.[22] Moreover, the volume of one's sensations is greatly expanded by increased perspicacity since the latter allows for the discovery of heretofore invisible aspects of one's milieu. Magnifying the density of experience slows the perceived passage of time because standard units of temporality are then bloated with an awareness of things that far exceeds what they contain under ordinary conditions.

The sensation of *protracted duration* ensues from the five elements in the foregoing sequence. Under such conditions, the individual perceives lived time to swell as standard units of temporality become saturated with the thick experience of subjective stimulus complexity. One has the impression that temporality is hugely, even infinitely magnified. The situation may seem endless, and the act-to-act flow of gestures may have the molasses-like viscosity that is evident in the slow-motion replay of a videotape. Typically, individuals reporting such episodes feel compelled to translate subjectivity into intersubjectivity. For instance, someone may say, "I know it only took a few minutes, but it felt like hours." Objectively "full" and objectively "empty" episodes have the same impact on the perceived passage of time. We can resolve this paradox by acknowledging that subjective involvement is crucial to our understanding of protracted duration.

Why is it that protracted duration emerges within the context of so-called empty intervals (e.g., solitary confinement) as well as intervals that are full of significant events (e.g., interpersonal violence)? The answer, of course, is that "empty" intervals are nothing of the sort. Such intervals are in fact filled with cognitive and emotional responses to one's predicament. A sharp transition from normal interaction to "empty" time ignites a preoccupation with aspects of self and situation that would have been overlooked in ordinary encounters. In particular, we often find that the person becomes more caught up in the rhythms of his or her own physiological existence.[23]

Engrossment, then, constitutes the underlying commonality for those situations that are, at first glance, strikingly divergent. It is a dimension of all social interaction; one is always more or less involved with one's activity (or inactivity). As Goffman has shown, the subjective involvement that participants bring to an encounter is of great importance.[24] His use of the term "engrossment" does not necessarily denote that the individual is enraptured by proceedings (although that is a possibility), but rather that he or she is paying selective attention to or is "spontaneously involved" with the immediate situation. Hence, one can be engrossed by circumstances that are distasteful or even horrible. In any event, situated engrossment is essential to our understanding of the process whereby protracted duration emerges from situations characterized by high or low levels of embedded activity.

Synchronicity

The empirical materials in chapter 3 are exclusively concerned with description of those circumstances that make for the experience of protracted duration. And, as we have seen, protracted duration arises in situations characterized by abnormally high or abnormally low levels of overt activity. Therefore, it is logical to assume that the "missing" circumstances are characterized by a moderate level of overt activity and that such circumstances bring about the normal experience of lived duration—that is, synchronicity. In short, we can apply the preceding analysis to our typical experience of lived duration (i.e., the absence of temporal anomalies) if we can assume that the preceding analysis describes only a relatively intense variation on the same process that operates under ordinary circumstances.

Most of the time, one does not experience protracted duration. Instead, the perceived passage of time is nearly synchronized with the standard temporal units of clocks and calendars. In other words, under normal conditions, ten minutes of time on the clock *feels* like approximately ten minutes of lived duration. How is it that we do not usually sense time passing quickly or slowly in commonplace encounters? What relationship between self and situation generates the usual conditions for rough synchronization between clock time and lived time? Given that subjectivity is implicated, how does it happen that the experience of eccentric temporality is atypical? We are acquainted with those circumstances that prompt the sensation of protracted duration. With certain modifications, the theoretical model that emerges from the analysis of those situations can also serve as an explanation for the normal experience of lived time.

Synchronicity is the typical form of temporal experience. Its predominance is neither natural nor accidental. Stanford Gregory points out that interpersonal relations "are not simply a set of independent behaviors; they are rather a set of temporally coordinated *interactions*."[25] In any society, orderly interaction requires the ongoing articulation of individual lines of conduct.[26] However, the problem of interpersonal coordination is especially acute among large urban populations.[27] Lewis and Weigert conclude that "modern industrialized and rationalized society can function only if most of its members follow a highly patterned and dependable daily round."[28] There is, then, regimentation of temporal experience in accordance with the time of clocks and calendars.

It is one thing to assert that synchronicity is necessary, and another to describe how it is accomplished. To understand how it is accomplished, students of temporality must grasp the connection between social order and subjectivity. Harold Garfinkel has demonstrated that "background expectancies" and social order are mutually constitutive.[29] Synchronicity is one of the background expectancies that make for orderly interaction. As such, it is a taken-for-granted and nearly unconscious aspect of temporal experience. Synchronicity is, nevertheless, a skill acquired in the course of primary socialization. Gradually, one learns not to cut encounters off too quickly or drag them out beyond their proper length. The regimentation of temporal experience is based upon one's awareness of social expectations.

We can elaborate on the formalization of our theoretical model by specifying five sequential factors that produce the sensation of synchronicity. To begin with, the typical form of the perceived passage of time arises from a situation that the individual finds familiar and comfortable. Within that context, the individual is pursuing a routine line of conduct, and the circumstances of that episode can be described as unproblematic. To be more precise, these situations are located in the middle ranges of two dimensions: from an objective standpoint, there cannot be too much or too little embedded activity; and from a subjective standpoint, the experience cannot be so customary that it is habitual, or so unusual that it is challenging. In brief, the situation is not characterized by extreme circumstances.

Consequently, those who find themselves in such situations experience only moderate emotional concern for the implications of what transpires. Goffman argues "that when individuals attend to any current situation, they face the question: 'What is it that's going on here?'"[30] In keeping with his formulation, we can postulate that, during one's waking life, one is nearly always experiencing some degree of concern for defining the situation correctly as it evolves through the various contingencies of interaction. We can refer to this horizon of concern as "framing anxiety." Typically, however, the individual's framing anxiety is tempered by the fact of routine conduct in a conventional setting. This reduces the salience of framing anxiety to such an extent that it may well be an unconscious facet of one's usual attention to situated proceedings. Indeed, Schutz and Luckmann note that the "natural attitude" of people in everyday life is the suspension of doubt, which is accompanied by taking for granted one's provisional understanding of events.[31]

When there is moderate salience for framing anxiety, there is equally moderate salience for cognitive involvement with one's circumstances. Of course, some degree of subjective involvement is nearly always requisite, if only to manage attention and one's availability for interaction. We are expected to avoid pratfalls, recognize our turn in conversations, and, in general, be alive to the many other exigencies that constitute ordinary wariness and sociability. Even so, this interactional regimen would be much more demanding than it is were it not for the fact that, under ordinary circumstances, the individual has no need to maximize situated involvement. Schutz and Luckmann have shown that the "pragmatic motive" is among the ruling principles of social interaction.[32] By this they mean that, typically, one's interests are restricted to the serviceable accomplishment of utilitarian tasks. Surprisingly little cognitive involvement is necessary if one assumes an instrumental attitude and aspires to nothing beyond a practical level of interactional competence.[33]

With subjective involvement of average intensity, only a temperate amount of stimulus complexity is generated by the individual's casual attention to self and situation. Both consciousness and the senses figure in the production of stimulus complexity because the latter reflects cognition and emotion as well as perception. What is more, due to the systematic quality of interpersonal relations, we gradually become accustomed to the modal quantity of experience (including thoughts, feelings, and perceptions) that is carried by each standard unit of temporality. In part, the modal quantity of stimulus complexity is governed by the social structure of activity: collective habits, schedules, and the like. But, given the socialization of subjectivity, consistency in the amount of experience per standard temporal unit is also conditioned by what Aaron Cicourel calls "interpretive procedures."[34] The individual uses interpretive procedures in an effort to sustain an approximate synchronization with others. This effort is called for because the etiquette of interaction dictates that encounters not be too short or too long in duration. Thus, synchronicity is only possible if, as Goffman observes, the individual modulates his or her subjective involvement with an eye toward conventional "rules for the management of engrossment."[35] Goffman adds that each of us must modulate involvement because "too much is one kind of delict; too little, another."[36] And, as we have seen, one cannot control subjective involvement without also controlling the stimulus complexity of the situation.

There is, then, a routine consistency in the density of experience per standard temporal unit. This consistency is an artifact of structuring processes that are both external and internal with respect to subjectivity. From the outside, the individual is entangled in a web of habits, schedules, calendars, seasons, and other socially defined regularities. From the inside, the individual draws upon the lessons of primary socialization to employ interpretive procedures in the self-conscious management of situated involvement. Unwritten rules prohibit overinvolvement as well as underinvolvement, thereby structuring a typical volume of interactional experience per standard temporal unit. In this regard, Goffman has revealed the elaborate codification of rules regulating subjectivity: "All frames involve expectations of a normative kind as to how deeply and fully the individual is to be carried into the activity organized by the frames."[37] Therefore, if temporal eccentricity is usually avoided, this is due to the socialization of subjectivity, and to the fact that societies provide recurrent things by which to be engrossed. The upshot is that the individual learns the ordinary correspondence between the flow of experience and the flow of standard temporal units. It is knowledge of this correspondence that enables one to translate experience into standard temporal units, and vice versa.

Our capacity for synchronicity results from the five elements in the foregoing sequence. Given ordinary conditions, the perceived passage of time is roughly synchronized with the intersubjective time that is measured by clocks and calendars. The normal adult in normal circumstances can estimate with fair accuracy the elapsed time of an interval by reflecting on the amount of experience he or she has processed during that interval. Young children have not mastered this skill, so it comes as no surprise that they are notorious for their impatience. Asked to "wait a minute," the young child quickly becomes frustrated because he or she has no familiarity with the volume of experience that typically fills a minute. Children receive explicit instruction in "how to tell time," and knowledge of standard temporal units is requisite for synchronization between subjective and intersubjective temporality. Nonetheless, the interpretive procedures that concern lived duration tend to be neither taught nor learned in formal fashion, but rather are gleaned implicitly from interaction subsequent to the basics of temporal socialization. Yet, as we have witnessed, even after socialization and mastery of these interpretive procedures, the vicissitudes of social interaction still confront the individual with great variety in the perceived passage of time. The emerging theory enables us to see the underlying unity beneath the heterogeneity in lived duration.

It is possible to summarize the relationship between synchronicity and protracted duration as a schematic figure.

The figure is a U-shaped curve, but one that improves on the model proposed by Hogan.[38] Like his model, Figure 1 shows that protracted duration arises in situations of high or low stimulus complexity.[39] However, unlike Hogan's model, Figure 1 also shows the correspondence between synchronicity and a moderate level of stimulus complexity. Still, Figure 1 does not encompass the full range of variation in the perceived passage of time. It has no place for those instances when time seems to have passed very quickly—that is, temporal compression.

Temporal Compression

Time flies. For centuries, this has been one of the stock phrases in Western civilization. But, on occasion, we are struck by the sense that time has passed even more quickly than is usually the case. This is to say that, in particular circumstances, it feels like much less time has elapsed than has actually been measured by the clock or calendar. Regardless of whether the relevant interval is ten hours or ten months, it seems to those of us in such circumstances that a much shorter length of time has gone by. Therefore, we can refer to this sensation as "temporal compression."[40]

One of the crucial insights to result from this study is that temporal compression is a facet of retrospection. In contrast to protracted duration and synchronicity, both of which are primarily phenomena of the present, temporal compression is uniquely associated with the past. Like Koestler's reaction as he circled the chalk marks on the

FIGURE 1: *Theoretical Relationship between Experience and Stimulus Complexity*

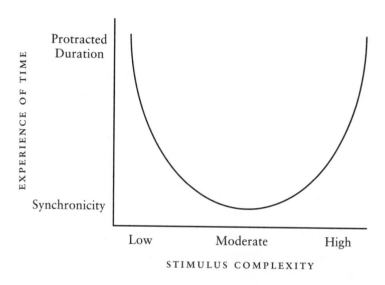

walls of his cell, its most familiar manifestation is the shocked look backward that is expressed in common questions: Where have the hours (days, months, or years) gone?[41] Protracted duration is experienced when the density of conscious information processing is high, and synchronicity is experienced when the density of conscious information processing is moderate. Given what we already know concerning protracted duration and synchronicity, it follows that we can extend the emerging theory by conceptualizing temporal compression as a product of situations in which standard units of temporality carry *less* conscious experience than is typically the case. My research has identified two factors that lower the density of experience per standard temporal unit: routine complexity and the loss of memory over time.

Routine Complexity

The writings of Mead and Dewey suggest that the flow of conscious experience is at its lowest ebb when the individual is engaged in habitual conduct. Mark Ashcraft and other cognitive psychologists refer to the subjective side of habitual conduct as "automaticity" or "automatic processing."[42] Moreover, they have noticed a recurrent trajectory in the learning of new skills:

Most mental processes begin at a conscious level. That is, they require conscious attention and processing resources at the outset. With extensive practice, however, it also appears that at least some of these mental processes can migrate toward the automatic end of the continuum. Thereafter, the mental activity can routinely happen at a very automatic, effortless level under normal circumstances, requiring little if any attention. If the situation changes dramatically, however, then some greater involvement of conscious processing may be required temporarily.[43]

Although the tendency is movement from conscious attention to automatic processing, once the latter degree of skill is achieved, there is oscillation between the two poles of this continuum, depending on whether the situation is relatively routine or relatively problematic. Habitual circumstances allow one to operate with largely automatic processing, but, as we saw earlier, challenging situations "capture attentional resources," because one "must revert back to more conscious control . . . under unusual circumstances."[44]

Habitual conduct and automatic processing play an important role in temporal compression, although these concepts have not been used in previous research on the perceived passage of time. If lived duration is perceived to slow (protracted duration) when the density of experience per standard temporal unit is greater than usual, then lived duration should be perceived to have quickened (temporal compression) when the density of experience per standard temporal unit is less than usual. This is exactly what transpires with habitual conduct and automatic processing because the individual is not directing his or her activity with conscious attention to the details of experience. Hence, we can extend our theoretical model of variation in the perceived passage of time by specifying five sequential factors that bring about the sensation of temporal compression.

First, the individual is pursuing a habitual line of conduct. The activity in question need not be inherently simple or easily managed, however. Indeed, from the perspective of society at large, it may be viewed as one of the more difficult things people do. With subjectivity at issue, we must remember that even highly complicated tasks can become, with adequate practice, forms of conduct one pursues with little or no conscious attention. In this sense, intrinsically challenging activities can be performed "mindlessly" when one is sufficiently prepared for their various contingencies.[45]

Second, one experiences a low level of emotional concern for one's capacity to deal with the situation. This does not mean that the individual is bored. As a matter of fact, the situation would be viewed as problematic were it not for extensive training or familiarity with its demands. The circumstances are complicated, but in a routine and thoroughly familiar fashion. In other words, the situation is characterized by what we can call "routine complexity." The upshot is that the well-prepared individual need not worry about his or her ability to handle what the situation has to offer.

Third, there is low cognitive involvement with self and situation. In the absence of problematic circumstances, the individual can act in nearly automatic fashion because habit and custom suffice as guidelines for behavior. There is, then, little or no need for thought on the part of those who, by dint of training or experience, can anticipate the various contingencies of their conduct. Working in circumstances that are familiar, if challenging, they are able to act without much premeditation. To put it another way, habitual conduct and automatic processing make it unnecessary to think about what one is to do next.

Fourth, there is an abnormally low level of stimulus complexity brought on by the near absence of attention to self and situation. A line of conduct is pursued in habitual or automatic fashion, and, as we have seen, the individual in question is neither emotionally nor cognitively involved with the proceedings. Still, it would be incorrect to assume that nothing is happening. There may be a great deal of overt activity, but, if it takes the form of routine complexity, it will not bring pressure to bear on attentional resources. Here again, it is well to remember that stimulus complexity is as much a function of subjectivity as it is of objectivity.

Fifth, the foregoing factors result in a lower than normal density of experience per standard temporal unit. Perception, emotion, and cognition are concepts that represent tacit mental activity. Such activity is unnecessary when one is operating on the basis of habitual conduct and automatic processing. Therefore, if we conceive of consciousness as a stream of experience—one that is composed of perception, emotion, and cognition—then it becomes evident that the flow of experience will be at low ebb in such circumstances. Or, returning to the imagery of gondola cars, we can say that, in effect, the seconds, minutes, hours, and so forth are carrying much smaller "loads" than they do under ordinary conditions. Whichever metaphor we choose, the point is that the volume of experience per standard temporal unit is much lower than usual.

The sensation of temporal compression is produced by the five factors in the foregoing sequence. In such circumstances, one has the impression that much less time has passed than has actually been measured by the clock or calendar. Furthermore, with this formulation, we are now in a position to resolve the second paradoxical aspect of lived duration: Why do some busy intervals seem to pass slowly while, in retrospect, others seem to have passed quickly? The answer becomes apparent when we realize that there is not just one kind of busy interval, but two: problematic complexity and routine complexity. With problematic complexity, the situated challenge demands that more than normal attentional resources be brought to bear on difficult circumstances. Protracted duration is the result. With routine complexity, the familiar task at hand requires that less than normal attentional resources be brought to bear on habitual circumstances. Temporal compression is the result. Thus, by conceptualizing two different kinds of busy intervals, we are able to integrate seemingly contradictory observations.

The Erosion of Episodic Memory

Routine complexity is the first of two factors that make for the sensation of temporal compression, and it is peculiar to those situations that are characterized by habitual conduct. The second factor that generates the sensation of temporal compression is the loss of memory over time, and this is the more common of the two. Routine complexity and automatic processing are most noticeable in our short-term memory of the near past (i.e., the last few minutes or hours), but forgetting is more evident in our long-term memory of the distant past (i.e., the last few days or months). And, as we will see, the responses of our subjects indicate that the further back you go, the more intensely is the effect of temporal compression felt. Consequently, it should come as no surprise that the older one is, the more quickly time seems to have passed.

As time goes by, there is little or no deterioration in semantic memory (knowledge) or procedural memory (technique). However, episodic memory concerns the details of social interaction in terms of specific events or experiences, and, generally speaking, this type of memory erodes with the passage of time.[46] Can anyone recall the detailed experiences of a day randomly chosen from the month before this one? Presumably, we ate meals, spoke with acquaintances, and worked at a variety of tasks, yet, in all likelihood, we have little or no recollection of exactly what happened on that day. Each individual's biography differs, but nearly everyone notes how the past seems to have gone by quickly. The homogeneity with which time erodes memory is reflected in the uniformity with which we perceive the passage of time: all of us must contend with the fact that time begins to erode episodic memory almost as soon as events have transpired.

As a general rule, the loss of memories over time reduces the quantity of experience that was carried by each standard unit of temporality. Hence, the past is constantly contracting in our memories, and the speed at which it seems to have transpired quickens as each quantum of experience is forgotten. The passage of time should be perceived to have accelerated when standard temporal units carry less experience than is ordinarily the case. Clearly, the erosion of episodic memory affects lived duration in a way that fits the emerging theory perfectly. The past is perceived to have passed quickly and, moreover, is perceived to have passed at an increasing rate, as the ongoing loss of memories erodes the density of remembered experience per standard temporal unit.

And here, finally, there is a resolution to the third paradoxical aspect of lived duration: How can the same period of time be experienced as passing slowly in the present but be remembered as having passed quickly in retrospect? As time passes, the details of situated experience fade from memory. In effect, the situation "contracts" in one's memory, and thereby seems to have passed more quickly than it did as measured by the clock or calendar. Episodic memory is oriented toward the recollection of activity. Situations in which there is an abnormally low level of overt activity generate the experience of protracted duration. But while these situations may seem to last forever as one endures them, little or nothing "happens," and they leave only a faint and waning residue in one's episodic memory.

As he endured solitary confinement, Koestler filled his time with intense contemplation of his predicament.[47] The duration of his sentence was burdened with the thick experience of perceptions, feelings, and thoughts, but not deeds or outward accomplishments. If memories and the relating of memories are both oriented toward tales of overt behavior, then what can Koestler say to himself or others about his imprisonment, save its objective length? He has few stories to tell, he has few stories to remember, and the ravages of time reduce still further what little there is to recall. Standard units of temporality that, at the time, were bloated with the intangibles of subjective involvement now appear almost empty in hindsight. The density of experience seems to diminish more, and the sentence seems to have elapsed more quickly, with each passing day.

The S-Shaped Model

Protracted duration is experienced when the density of conscious information processing is high; synchronicity is expe-

FIGURE 2: *Theoretical Relationship between Experience
of Time and Intensity of Conscious
Information Processing*

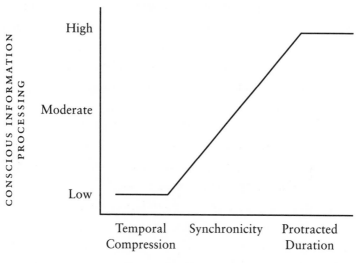

EXPERIENCE OF TIME

rienced when the density of conscious information processing is moderate; temporal compression is experienced when
the density of conscious information processing is low.
Once again, it is possible to summarize this relationship as a
schematic figure. Figure 2 is an S-shaped curve, and it represents a significant advance over previous models of variation in the perceived passage of time. Unlike earlier models,
the S-shaped curve does encompass the full range of variation in lived duration. As a result, the formulation of this
model resolves each of the three paradoxical aspects of
temporal experience. Moreover, this model explicates the

conditions that give rise to each of the three fundamental forms of temporal experience: protracted duration, synchronicity, and temporal compression.

The full complexity of lived duration is not apparent unless one leaves the laboratory in search of the tales people tell about their temporal experiences in everyday life. Their paradoxical stories make it clear that, if there is to be accurate conceptualization of variation in lived time, self and subjectivity must be given the same weight as situation and objectivity. What is more, the self should not be viewed as merely a passive respondent to environmental stimuli, but as an active agent, modulating selective involvement with the flux of inner and outer experiences.

These principles form the foundation for the theoretical model presented in this chapter. It is a formal model of generic processes that condition the experience of lived duration. The model has been formulated through inductive theory construction, using empirical materials drawn from the triangulation of qualitative methods. Its central proposition is that variation in the perceived passage of time reflects variation in the intensity of conscious information processing per standard temporal unit. This model having been generated, the next step involves deductive derivation of hypotheses that can be examined against the backdrop of new empirical materials. In this fashion, progress can continue toward the goal of comprehending variation in the perceived passage of time.

[5]

Temporal Compression

T HERE IS A common misconception that portrays research as linear movement from ignorance to insight. The actual practice of research is represented more accurately by the circular model formulated by Walter Wallace.[1] In his conception, observations lead to empirical generalizations which, in turn, lead to the construction of explanatory theory. These steps toward increasing abstraction constitute the inductive side of what Earl Babbie calls the "wheel of science."[2] But theory construction does not end the process, because the resulting theory will have empirical implications. Thus, hypotheses are derived from the theory, and the empirical implications of these hypotheses are examined in light of new observations. These steps toward decreasing abstraction constitute the deductive side of the circular model. And, of course, the new observations lead to confirmation or modification of empirical generalizations which, in turn, lead to confirmation or modification of one's theory. There is, then, no point at which the process ends.

The circular model can be misconstrued, however, because it suggests that researchers go round and round without ever getting anywhere. How can a wheel represent work that advances our theoretical understanding? In this regard, perhaps the spiral is a more accurate image. It preserves the circular process depicted by Wallace, but adds a sense of progress with its rising structure. The practice of research may be circular, but our resulting knowledge can be said to spiral upward as theoretical understanding grows in its capacity to account for a wider and wider range of variation. Theory construction produces a system of interlocking propositions which, taken together, describe and explain the phenomenon in question. Therefore, we can speak of advancement in our theoretical understanding whenever one theory is supplanted by another because the new theory is capable of accounting for variation left unexplained by its predecessor.

Description and explanation do not exhaust the functions of theory. Norman Denzin adds that theory "furnishes the basis for the prediction of events as yet unobserved."[3] This chapter marks a transition from the construction of theory to the evaluation of theory. Of course, in a sense, evaluation was already at issue in the preceding chapter. During the inductive process of theory construction, the emerging model was evaluated in terms of its capacity to account for the full range of variation in the perceived passage of time. Now, however, evaluation will be the explicit focus of attention. The theory will be evaluated in terms of its capacity to predict the empirical form of temporal experiences; moreover, these temporal experiences will be specified by hypotheses deduced from the theoretical model.

According to King, Keohane, and Verba, the introduction of new empirical materials is crucial to this transition: "[E]vidence should be sought for hypotheses in data other than that in which they were generated."[4] Consequently, this chapter will present new empirical materials. Some of them were generated by me, and the relevance of these data for the evaluation of my theory is intentional; some of them were generated by other scholars in the course of their own research, and the relevance of these data for the evaluation of my theory is unintentional. My confidence in the findings is strengthened by the corroboration of the intentional and unintentional evidence.

Hypotheses

In an earlier chapter, the concept of protracted duration was generated inductively through the use of qualitative methods. That component of the theoretical model is already grounded in empirical materials. However, I have extended the theory to include the concept of temporal compression by identifying two factors that should reduce the density of experience per standard temporal unit: routine complexity and the erosion of episodic memory. These concepts are plausible extensions of the theoretical model, but they are, as yet, untried against empirical materials. What is more, in contrast to my previous use of inductive logic and qualitative methods, these components of the theory representing temporal compression have deductive and quantitative implications that lend themselves to straightforward statistical tests. In short, systematic empiricism requires the use of multiple methods.

A call for the use of multiple methods has become a commonplace feature of methodological texts.[5] However, any thorough review of the research literature reveals that the use of multiple methods is encouraged more often than it is practiced. And, of course, it is even more unusual to find the integration of quantitative and qualitative methods within the context of a single study. Nonetheless, with King, Keohane, and Verba, I assume that the differences between quantitative and qualitative research are stylistic, not epistemological.[6] Furthermore, if we assume that the logic of scientific method provides the underpinning for both types of inquiry, then the use of quantitative and qualitative methods is in keeping with a call for the use of multiple methods.

I have derived two hypotheses from the theoretical model, both of which concern the relationship between memory and temporal compression. First, I have hypothesized that time will be perceived as having passed more quickly the further back one is asked to remember. According to my theory, the erosion of episodic memory lowers the density of remembered experience per standard temporal unit. Consequently, last year should be perceived as having passed more quickly than last month, and last month should be perceived as having passed more quickly than yesterday. Second, I have hypothesized that, on average, my subjects will not report protracted duration when asked how they perceive time to have passed yesterday, last month, and last year. Again, if the erosion of episodic memory lowers the density of remembered experience, then we should find that the average response is at least 3.0 on a scale where 1 = Very Slowly, 2 = Slowly, 3 = Normally, 4 = Quickly, and 5 = Very Quickly.

Neither of these hypotheses addresses automatic processing, because empirical examination of that concept requires methods that are quite different from those used to examine the erosion of episodic memory. In cognitive psychology, the operational definition of automatic processing involves the use of interference designs. Subjects are said to engage in automatic processing when their performance of a primary task is not degraded by simultaneous performance of a secondary task which has been shown to demand attentional resources. Hence, automatic processing does not lend itself to direct measure of recollection by pencil-and-paper instruments. It is a concept that refers to activity one pursues with little or no conscious involvement, and there is, consequently, little or nothing to remember. However, the relationship between automatic processing and temporal compression has been addressed, albeit unintentionally, by the research of other scholars. Later in this chapter, I will examine the implications of their findings for my theoretical model.

Methods

Sample

The age of my subjects was an important consideration. First, there was the possibility that the hypothesized relationship between memory and the perceived passage of time

might be conditioned by the subject's age. Second, I anticipated that age might have its own effect on the perception
of time. Fortunately, there were three undergraduate programs on our campus, each of them serving a very different
constituency in terms of age. The subjects were 366 students drawn from courses in those three programs.[7] The
first subgroup (N = 122) consisted of traditional students
enrolled in the residential program. Their mean age was
19.8, with 46.8 percent male and 53.2 percent female. The
second subgroup (N = 122) consisted of nontraditional students enrolled in the continuing education program. Their
mean age was 38.4, with 42.6 percent male and 57.4 percent female. The third subgroup (N = 122) consisted of students enrolled in the Elderhostel program which brings
older people to campus for brief noncredit courses. Their
mean age was 71.2, with 38.3 percent male and 61.7 percent female. I selected courses with similar enrollments, and
accepted only the necessary number of volunteers, thereby
generating an equal number of subjects in each of three subgroups: young, middle-aged, and elderly.[8]

Measurement

In measuring my subjects' perception of the passage of
time, I used three Likert-type scales where 1 = Very Slowly,
2 = Slowly, 3 = Normally, 4 = Quickly, and 5 = Very
Quickly. I measured the effects of memory by asking the
subjects to consider three different periods of their lives:
yesterday, last month, and last year. The data were collected
during the first week of March 1990. Subjects were in-

structed to define the phrase "last month" as February 1990, and they were instructed to define the phrase "last year" as 1989. My questionnaire asked the subjects to take a few moments and think about yesterday. They were asked to think about their activities as well as their experiences. Then they were asked to select the number most accurately indicating how time seemed to have passed yesterday. This same format was used to measure their perception of how time seemed to have passed last month and last year. I hypothesized that the subjects would remember less of their experiences last year than of last month, and less of their experiences last month than of yesterday.[9]

Design

The dependent variable is the subject's perception of the passage of time, and it consists of the subject's selections of one number from each of the three Likert-type scales. There are two independent variables: (1) the subject's age (young, middle-aged, or elderly); and (2) the subject's memory of his or her experiences yesterday, last month, and last year. Each of the 366 subjects was in one of three age groups, and each subject contributed 3 of the 1,098 scores on the Likert-type scales. The subject's age is a between-groups factor, and the subject's memory of yesterday, last month, and last year is a repeated-measures or within-subjects factor. Thus, I used a two-factor mixed design for the analysis of variance.[10] I tested the main effects of age and memory as well as the possibility of interaction effects using the MANOVA program from SPSS.[11] In addition, I examined the differences between

TABLE 2

Analysis of Variance for Temporal Compression

Source	Sum of Squares	Degrees of Freedom	Mean Square	F	Probability
Between Subjects					
Age	12.60	2	6.30	5.74	.004
Error	398.14	363	1.10		
Within Subjects					
Memory	107.66	2	53.83	94.38	.000
Age x Memory	.92	4	.23	.40	.806
Error	414.08	726	.57		
Total	933.40	1,097			

pairs of means using two-tailed protected t tests—uncorrelated for the between-groups factor, correlated for the within-groups factor.

Results

Table 2 presents a summary of the analysis of variance. The results show strong support for my hypotheses. Both of the expected main effects are statistically significant. Clearly, the subject's perception of the passage of time is affected by the subject's memory of his or her experiences yesterday, last month, and last year. Just as clearly, the subject's age (young, middle-aged, or elderly) is a factor shaping his or her perception of the passage of time. In addition, Table 2 shows that there is no evidence of interaction between the two factors of age and memory.

Table 3 presents the means and standard deviations for each subgroup in response to my three questions concerning the perceived passage of time. This table allows us to examine differences among the means across categories of the independent variables, thereby enabling us to give a more detailed analysis of the two main effects.

The Effects of Memory

In accordance with the theory concerning temporal compression, subjects perceived time to have passed more quickly the further back in time they were asked to remember. I had hypothesized that the means for the perception of time yesterday, last month, and last year would be in reverse rank order to their objective length. Table 3

TABLE 3

Means and Standard Deviations of Responses to Questions on Perceived Passage of Time by Age of Respondent

	Yesterday		Last Month		Last Year	
	Mean[a]	SD	Mean	SD	Mean	SD
Young (N = 122)	3.402	.888	3.836	.965	4.131	.953
Middle-Aged (N = 122)	3.631	.947	4.115	.718	4.328	.673
Elderly (N = 122)	3.344	.851	3.885	.883	4.189	.846
Full Sample (N = 366)	3.459	.902	3.945	.868	4.216	.834

[a]1 = Very Slowly, 5 = Very Quickly.

shows that, as expected, the mean for yesterday is lower than the mean for last month, and the mean for last month is lower than the mean for last year in the full sample as well as all three age groups. This, despite the fact that, from the objective standpoint of standard temporal units, yesterday is approximately one-thirtieth of last month, and last month is approximately one-twelfth of last year. Twelve *t* tests are possible for differences between pairs of means: yesterday by last month, yesterday by last year, and last month by last year for three age groups and the full sample. All twelve *t* tests are statistically significant with $p < .008$ in each case.

The theory asserts that the experience of protracted duration is a response to subjective involvement in unusual circumstances. By definition, such episodes are rare, and the memory of them erodes as time passes. Subjects who wanted to characterize their perception of time yesterday, last month, or last year in terms of protracted duration would have selected a 1 (Very Slowly) or a 2 (Slowly) from the Likert-type scales. Therefore, I hypothesized that none of the means for the three scales would be lower than 3.0. Table 3 confirms my prediction. In fact, subjects selected a 1 or a 2 for only 8.4 percent of their 1,098 responses: 52 times for yesterday, 26 times for last month, and 14 times for last year. Furthermore, each of the latter two frequencies is roughly one half the size of the preceding one. The descending order of these frequencies highlights the way in which the erosion of episodic memory affects the perceived passage of time.

The Effects of Age

The findings with respect to age were somewhat serendipitous. While I anticipated that age might have a direct effect, I did not expect it to be so strong and systematic. The means for the middle-aged subgroup are higher than the means for the other two age groups in each of the three time periods. This pattern indicates that the middle-aged subgroup consistently perceives time to have passed more quickly than do the other age groups. There are no statistically significant differences between the means of the young and the elderly subgroups; all of the statistically significant differences are found when contrasting the middle-aged subgroup with one of the other two age groups.

Looking first at the means for yesterday, we find that the mean for the middle-aged subgroup is significantly different than that of the young ($p = .052$) as well as that of the elderly ($p = .013$). Similarly, when we examine the means for last month, we find that the mean for the middle-aged subgroup is significantly different than that of the young ($p = .011$) as well as that of the elderly ($p = .027$). It is only in the contrasts for last year that the differences between the means for the middle-aged and the means for the other two subgroups are not quite significant ($p = .064$ for the young and $p = .156$ for the elderly). Here, presumably, the powerful effect of looking so far back in time begins to make for some homogeneity in the way the three age groups perceive the passage of time.

In computing the F value for the main effects of age, the MANOVA program creates an index for the perception of time by adding the subject's scores on the three scales (yesterday, last month, and last year). This index serves as the dependent variable in a one-way analysis of variance, and the program computes a mean for this index within each of the three age groups (young, middle-aged, and elderly). The differences among the three means for the three age groups become even more distinct when this index is used as the basis for the protected *t* tests. Again, there is no significant difference between the means for the young and the elderly ($p = .838$). Moreover, as before, the mean for the middle-aged subgroup is significantly different than that of the young ($p = .002$) and the elderly ($p = .004$).

I suspect that the age structure of my sample serves as an indirect indicator of variation in the degree of routine complexity which characterizes different stages in the life cycle.[12] Certainly there is considerable evidence supporting this interpretation. Carol Ryff observes that middle-aged people experience greater routine complexity in their daily lives than do the young or elderly.[13] Her data dovetail with those of John Robinson who finds that "it is adults aged 35 to 54 who are most likely to report feeling rushed."[14] Bertram Cohler notes that, with adulthood, one usually acquires "a complex role portfolio . . . including marriage, parenthood, and work."[15] Likewise, David Gutmann describes "parenthood" as "a period of chronic emergency."[16] His phrase "chronic emergency" is consistent with what I have termed "routine complexity."

In contrast, research by the Carnegie Corporation of New York shows that approximately "40 percent of adolescents' waking hours are discretionary—not committed to other activities (such as eating, school, homework, chores, or working for pay)."[17] At the other end of the age spectrum, the elderly must contend with what Irving Rosow calls "role loss" due to the departure of their children as well as their own illness, retirement, and widowhood.[18] David Gutmann refers to this process as "disengagement."[19] Given the logic of my theory, if middle-aged people deal with more routine complexity, then we should expect them to perceive time as having passed more quickly. Still, this is a post hoc interpretation, and verification of that portion of the theory concerning routine complexity must await the more direct examination of its effects in the balance of this chapter.

In sum, the effects of memory on the perception of time seem to be more general than do the effects of age. While age generates temporal compression only among the middle-aged, the erosion of episodic memory generates temporal compression among all age groups.

A Brief Excursus on Synchronicity

The focus of this chapter is that portion of the theory concerning temporal compression. However, in post hoc fashion, a variation on the second hypothesis enables us to address that portion of the theory concerning synchronicity.

The theoretical model asserts that synchronicity is the normal experience of lived duration. Moreover, the number 3 was used on the Likert-type scales to stand for the normal passage of time. Since yesterday represents so little erosion of episodic memory, we should not expect much in the way of temporal compression. And, as indicated by the second hypothesis, we should not expect protracted duration either. Therefore, given an assumption that synchronicity is the normal experience of lived duration, a third hypothesis might have been worded as follows: the mean response for yesterday will be greater than 3.0 (no protracted duration) and less than 3.5 (little temporal compression).

Returning to Table 3, we find that the means for yesterday are 3.402, 3.631, 3.344, and 3.459. Only the mean for the middle-aged subgroup lies outside the specified range, but this slight divergence may be attributable to the relatively high amount of routine complexity which is confronted by those who are middle-aged. In any event, this variation on the second hypothesis provides support for that portion of the theory concerning synchronicity.

Temporal Compression in Restaurant Kitchens

Thus far, the research reported in this chapter demonstrates that the experience of temporal compression can be produced by the erosion of episodic memory. In addition, there is evidence to suggest that the experience of temporal compression can be prompted by routine complexity and automatic processing. However, the age structure of my sample is not a straightforward measure of routine complexity, and the analytical distance between these two concepts leaves the foregoing interpretation somewhat speculative.

Happily, the relationship between routine complexity and temporal experience has been examined more directly in two other studies. One is an article by Gary Alan Fine, entitled "Organizational Time: Temporal Demands and the Experience of Work in Restaurant Kitchens."[20] The other is a chapter by Mihaly Csikszentmihalyi, entitled "Enjoying Work: Surgery."[21] In contrast to my own research on temporal compression, both of these studies are based on qualitative data.[22] Hence, once again, the findings are grounded in the use of multiple methods. What is more, since neither of these scholars set out to assess my theoretical model, their findings are untainted by that tacit prejudice which the author of a theory brings to its evaluation.

Fine's research reveals some of the ways in which customer expectations condition how chefs experience their work in restaurant kitchens. Of particular relevance are his findings with respect to the "rush"—those busiest, but relatively brief intervals during which chefs must cook for a great many people. Like most workers, the chefs "prefer evenings when they discover that it is much later than they thought."[23] Put differently, it is desirable that their time in the kitchen seem to have passed quickly. However, temporal compression is experienced only if the rush goes "smoothly."[24] Now it may seem incongruous to describe the rush as going smoothly, but, while the rush may seem chaotic to one who lacks the requisite skills, the chef is actually performing habitual tasks within the context of an utterly routine situation. Consequently, Fine provides this summary of his interviews: "Several cooks remarked that their reactions are 'automatic' in that they do not consciously plan or control their emotions or the behavior resulting from the demands made of them."[25] It does not seem coincidental that variations on the words *automatic* and *unconscious* appear so frequently.[26] Fine and his informants are trying to articulate the temporal compression that occurs during the rush when everything goes smoothly.

Nevertheless, things can go wrong, and one of Fine's chefs observes that this possibility creates two kinds of rush: "'When you're all set and you're ready for it, it can be great. When things are happening that aren't supposed to, it can be a nightmare.'"[27] These two kinds of rush have very different impacts on the perceived passage of time. With the routine rush, lived duration collapses in on itself, as habitual conduct and automatic processing make for the retrospective sensation of temporal compression. On such occasions, chefs ask themselves where the night has gone. With the problematic rush, unanticipated difficulties and resulting emotional concern prompt an extraordinary amount of conscious attention to self and situation. This increased subjective involvement generates stimulus complexity, bloating standard temporal units with a density of experience that far surpasses their usual volume. On such occasions, the night seems like it will never end.[28]

As we saw in chapter 2, one of the paradoxical aspects of lived duration is that time is perceived to pass slowly in situations with abnormally high or abnormally low levels of overt activity. Therefore, it comes as no surprise when Fine reports that chefs perceive time to pass slowly "when there are either too many external demands or not enough."[29] This paradox was resolved in chapter 4 by positing that such intervals are equally full of subjective involvement with self and situation. Fine's observations are quite predictable in light of this conceptualization. And again, as we saw in chapter 2, a second paradoxical aspect of lived duration is that some busy intervals are perceived as passing slowly while others are perceived as having passed quickly. This paradox was resolved in chapter 4 by positing two kinds of busy intervals: routine complexity, which is characterized by temporal compression, and problematic complexity, which is characterized by protracted duration. This distinction is mirrored perfectly in the aforementioned statement where one of Fine's chefs describes two kinds of rush.

Fine's study would not have been more appropriate had I designed it myself. He had no intention of assessing that portion of my theory concerning routine complexity and automatic processing, yet his findings are exactly what my theoretical model would have predicted. In short, his study is tantamount to substantiation of my theory within the context of new data.

Temporal Compression during Surgery

Csikszentmihalyi has formulated a theoretical model of those circumstances that make for an enjoyable experience.[30] He uses the term "flow" in reference to enjoyable experiences, and he argues that a sense of flow results when one's skills are a good match for the demands of the immediate situation. Boredom is generated when one's skills are greater than what the situation demands, while anxiety is generated when one's skills are less than what the situation demands.

Csikszentmihalyi finds that surgeons enjoy their work. In fact, they claim that they would want to perform surgery even if it were not associated with lavish salaries and high prestige. Moreover, in keeping with his theoretical model, he finds that surgery is most enjoyable when the surgeon is challenged but not overwhelmed by the task at hand: "There seems to be a clear distinction between 'very routine' cases, which are perceived as boring, and 'routine' situations, which provide a relaxing if not completely absorbing involvement with the activity."[31] Yet, despite training and experience, surgical procedures can go awry, and Csikszentmihalyi notes that this alters the atmosphere in the operating theater: "The state of flow gives way to tenseness and then to anxiety if something goes wrong, if the operation changes from challenging to problematical."[32] With words reminiscent of Fine's study, one of the surgeons reports that the "'smoothness of the operation disappears'" in problematic situations.[33] To be sure, the chef and the surgeon provide different services. Ideally, however, both operate under challenging circumstances, and both want things to go "smoothly."

Those who cook in restaurant kitchens must contend with variation in the challenge of their work. On some nights, there are not enough customers; on other nights, there are too many customers; and on still other nights, there are just enough customers to require that one cook with all of the skill at one's disposal. In analogous fashion, we can conceive of three kinds of surgery: procedures that are unchallenging, those that are challenging, and those that are problematic. The surgeons are bored when procedures are unchallenging, and they are anxious when procedures are problematic. Either way, the increased allocation of attentional resources to self and situation amplifies the density of experience per standard temporal unit, which, in turn, makes for protracted duration—the perception that time is passing slowly. Here, we see clear corroboration for the curvilinear model in chapter 4. In contrast, when the procedure is challenging but unproblematic—a situation I have described as "routine complexity"—we can anticipate that the surgeons will be able to act in an unthinking fashion by dint of training and experience. These circumstances lower the density of conscious information processing per standard temporal unit, and my theoretical model would predict temporal compression—the perception that time has passed quickly. This prediction is confirmed by Csikszentmihalyi's findings: "Seventeen of the [twenty-one] surgeons interviewed said that they are either unaware of time or, more commonly, that it seems to pass much faster."[34]

Further substantiation is provided by the informants' use of variations on the word "automatic." One of Csikszentmihalyi's surgeons offers the following explanation for his ability to deal with routine complexity: "You are trained—you're already prepared. Your training makes you confident and you can [act] 'automatically.'"[35] Previously, we have seen that variations on the word "automatic" are crucial to the way in which chefs describe their cooking during the rush in restaurant kitchens. What is more, variations on the same term appear elsewhere in Csikszentmihalyi's book when he reports on his interviews with rock climbers:

"It's like when I was talking about things becoming 'automatic' . . . almost like an egoless thing in a way—somehow the right thing is done without . . . thinking about it. . . ."

"When things are going poorly, you start thinking about yourself. When things go well, you do things automatically without thinking."[36]

If variations on the word "automatic" are volunteered by those who are trying to describe routine complexity in cooking, surgery, and rock climbing, then we have far-reaching evidence in support of the proposition that automatic processing forms one basis for the experience of temporal compression.[37]

Coming Full Circle

Observation and interpretation are alternating phases of the research act. One has to do with the things of this world; the other, with our understanding of those things. There is a sense, then, in which the research act involves a kind of travel to and from the things themselves. Thus, Herbert Blumer provides the epistemological framework for this project:

I shall begin with the redundant assertion that an empirical science presupposes the existence of an empirical world. Such an empirical world exists as something available for observation, study, and analysis. . . . This empirical world must forever be the central point of concern. It is the point of departure and the point of return in the case of empirical science. It is the testing ground for any assertions made about the empirical world.[38]

My study of lived duration began in the empirical world. People who find themselves in unusual circumstances report systematic alterations in the perceived passage of time. In chapter 3, such reports furnished evidence for a descriptive analysis aimed at grasping the underlying commonalities among situations which seemed, at first glance, strikingly divergent. In chapter 4, we left the empirical world so as to formulate a theoretical model that accounts for the full range of variation in the perceived passage of time. In chapter 5, we have returned to the empirical world to evaluate how well predictions derived from that theoretical model fit with the real experiences of people in everyday life. And, of course, these findings provide impetus for further research. It is the oscillation between observation and interpretation that makes for spiral progress toward greater understanding of temporal experience.

[6]

Conclusion

L IKE ALL FORMS of human activity, research and writing are grist for the mill of temporal reflection. So it is with something of the usual shock (and more than a little dismay) that I look back over the work of this project and ask myself where the time has gone. And, of course, duration is not the only concern; timeliness is also at issue. "'The time has come,' the Walrus said," but an ending of a different sort is called for here.[1] By way of concluding, then, I want to address the theoretical implications of my findings and discuss some directions for future research.

Theoretical Implications

Randall Collins has argued that George Herbert Mead "developed a sociological theory of mind," albeit one that is in need of revision and further development.[2] To that end, Collins has proposed a theory of "interaction ritual chains" which extends Mead's model by elaborating on the latter's simplistic view of role taking:

One might say there is a continuum from tentative and ambiguous role-taking at one end, to completely certain role-taking at the other. A key ingredient of ritual interaction is that the role-taking is at the high certainty end of this continuum.[3]

Implicit in this statement is the assumption that the individual will need to devote more or less attentional resources to interaction depending on the degree to which the immediate situation is problematic.

Mead and Dewey never confront the full implications of this assumption for their concepts of mind and self, but these forerunners of symbolic interactionism suggest that there is variation in the volume of subjective experience, and that this variation reflects variation in the nature of one's circumstances. The phenomenologists—Heidegger, Husserl, and Schutz—direct our analysis to internal time consciousness, and they demonstrate that temporality varies across different realms of being. From Goffman's microstructuralism we learn that subjective involvement with self and situation is conditioned by social conventions. According to the ethnomethodologists, Garfinkel and Cicourel, these social conventions constitute background expectancies enacted in largely tacit fashion through interpretive procedures. In short, my theoretical model integrates elements of symbolic interactionism, phenomenology, microstructuralism, and ethnomethodology. These writings serve as the foundation for a theory that explains variation in the perceived passage of time. All manner of objections can be raised concerning my syncretic approach to what are, after all, apples and oranges. But where those concerned with classification see divergence, those seeking discovery see convergence and a platform for continued development. Thus, I do not view these writings as incompatible texts, but rather as disparate stepping stones which, when properly aligned, allow us to climb a little higher and see a little further.

A theory that is meant to account for variation in the perceived passage of time must take into consideration the interplay of different levels of consciousness. The theory that has emerged from my research recognizes that the density of conscious experience is a crucial element conditioning one's perception of the passage of time. In effect, I have argued that protracted duration occurs when conscious information processing is high, synchronicity occurs when conscious information processing is moderate, and temporal compression occurs when conscious information processing is low. My theoretical model for variation in the perceived passage of time acknowledges the interplay of different levels of consciousness. The model asserts that the intensity with which one directs attentional resources to the situation at hand varies according to where that situation falls along a continuum from abnormally repetitive activity to abnormally problematic activity. Variation in the nature of the social situation, and its implications for self, govern variation in the level of conscious information processing. In turn, variation in the level of conscious information processing is reflected by the density of experience per standard temporal unit. Ultimately, it is the density of experience per standard temporal unit that determines variation in the perceived passage of time.

With this theoretical model, we are in a position to subsume, at a higher level of abstraction, more particular observations concerning the perceived passage of time. Denzin notes that "Intense emotionality appears to stop time."[4] Katz contends that rage "magnifies the most minute details," creating the potential for "an endless present."[5] And Charmaz describes "the dragging time of pain and suffering."[6] In each case, these findings can be recognized as specific manifestations of those factors that generate the experience of protracted duration. Likewise, Fine and Csikszentmihalyi discover that chefs and surgeons who deal with routine complexity are surprised to learn that much more time has elapsed than they had anticipated.[7] In both cases, these observations can be recognized as specific examples of those factors that make for the experience of temporal compression. Thus, these and other empirical statements provide corroboration for a theoretical model that transcends the particulars of this or that research setting, thereby providing us with a broader and deeper understanding of the ways in which subjective and situated processes condition the perceived passage of time.

This theoretical model is predicated upon a fundamental observation: experience is shaped by variation in the magnitude of involvement. In any given situation, more or less of it will be deemed appropriate by society, and, in the manner of a self-fulfilling prophecy, the conventional level of involvement is typically viewed as a natural or inevitable facet of the occasion. Still, one's actual level of involvement is frequently conditioned by one's circumstances or willed by one's intentions. In turn, experience provides the informational foundation for mind, self, thought, emotion, interaction, deviance, and the definition of the situation, not to mention variation in the perceived passage of time. Goffman points out that a certain level of involvement can transform participants "into worthy antagonists in spite of the triviality of the game, great differences in social status, and the patent claims of other realities."[8] Borrowing Bentham's concept of "deep play" for his famous essay, "Notes on the Balinese Cockfight," Clifford Geertz shows that, at one level of involvement, the cocks are just birds (and exploited birds at that), but at another level of involvement, they are a warrior aristocracy engaged in thrilling combat— combat so invested with meaning that the wagering of vast sums of money makes sense (and adds another level of consequentiality to the proceedings).[9] Different levels of involvement create different worlds. Our inquiries must be capable of navigating the various levels of involvement if we are to understand the worlds of experience they bring into being.

Human experience—including, of course, temporality— is always embedded within the dynamics of social interaction. According to George Herbert Mead, "We should undertake to state the experience of the individual just as far as we can in terms of the conditions under which it arises."[10] When we follow Mead's advice, it appears that variation in social interaction makes for variation in human involvement (including both its cognitive and affective forms), and that variation in human involvement produces variation in temporal experience. Inevitably, then, the study of human experience implicates the study of social interaction, and vice versa. What is more, human experience is always at issue in the social construction of reality.[11] Different kinds of interaction generate different modes of experience, and different modes of experience represent different realms of being. Indeed, even anomalous experiences are recognized and interpreted against the backdrop of social conventions. Thus, this project serves as a model for an empirical, naturalistic, interactionist, and constructionist approach to the sociology of subjectivity—an approach that integrates the sociology of emotions with cognitive sociology.[12]

A sociology of subjectivity is centrally concerned with how the mind is conditioned by social interaction. Lindesmith, Strauss, and Denzin depict the mind as an interpretive entity that stands astride two intersecting streams of information.[13] The mind must interpret information from its external environment (i.e., one's situation or circumstances) as well as its internal environment (i.e., self-consciousness, cognition, emotion, and the experienced body). With his famous metaphor, William James describes these intertwining processes of interpretation as the "*stream . . . of consciousness.*"[14] This study demonstrates the usefulness of extending his metaphor so that we can consider the *volume* or *density* of this flow of information and interpretation. As we have seen, the density of conscious information processing is indispensable to our understanding of variation in the perceived passage of time, and, in turn, the density of conscious information processing seems to vary with the level of one's involvement in the immediate situation. Social conventions prescribe a certain level of involvement for each occasion, but particular circumstances intensify one's involvement, and, of course, individuals can conjure a particular level of involvement at will (a subject to which we must return later). In light of social and personal efforts to control the volume of experience, it would appear that the stream of consciousness is a heavily managed flow.

The sociological version of social psychology began with efforts to grasp the difference between conduct that is organized by habit and custom, and conduct that is improvised in response to problematic circumstances. All too often, this distinction has been conceptualized as a dichotomy, but the findings of this study show what there is to gain by conceiving of it as a continuum. What happens if we apply this same logic to other concepts, such as mind, self, cognition, and emotion? What happens if we assume that, like the perceived passage of time, they reflect oscillation in the flow of information and interpretation (i.e., experience)—oscillation that is modulated by involvement and interaction? When conceived in this fashion, it becomes evident that there are advantages to thinking about the assorted aspects of subjectivity in terms of variation.

Directions for Future Research

Having examined, albeit briefly, some of the theoretical implications of the S-shaped model, let us turn now, by way of conclusion, to address its implications for further research.

There is, of course, a sense in which social investigation always produces more questions than answers, and, with respect to the S-shaped model, the primary question concerns its cross-cultural validity: Is this theoretical model applicable to cultures other than our own? Thus, the anthropological literature on temporality must serve as something of a springboard for this inquiry.[15]

The Cross-Cultural Question

Does the S-shaped model for variation in the perceived passage of time depend on a cultural context that is organized around clocks and calendars? If the answer to this question is "no," then the theory that emerges from this study can help us understand the temporal experiences of people in technologically less advanced societies. As it stands, however, the anthropological literature does not encourage much optimism. Beginning with Martin Nilsson's book, *Primitive Time-Reckoning*, its dominant theme has been an emphasis on how temporality differs across cultures.[16] Irving Hallowell introduced the reader to his research on the Saulteaux Indians with a prototypical statement: "Like other cultural phenomena, temporal frames of reference vary profoundly from society to society."[17] Subsequent research was concerned with "time-reckoning" among the Nuer and "concepts of time" among the Tiv.[18] A great deal of cross-cultural continuity can be observed in their findings (mostly having to do with cyclical conceptions of time rooted in naturally recurring phenomena, such as the alternation of day and night, seasons, and phases of the moon), but these dimensions of temporality were thought to be fundamentally different from those that are characteristic of our own culture.

The anthropological emphasis on temporal relativism continued with the linguistic work of Benjamin Lee Whorf. Based upon his research, he argues that "the Hopi language contains no reference to 'time,' either explicit or implicit."[19] He concludes that some languages are simply "better equipped to deal with" certain phenomena (e.g., temporality) than are others.[20] His attention to linguistic resources encouraged the cross-cultural study of terminology and conceptualization. Murray Wax examined "the Pawnee conception of time," while T. O. Beidelman described "how the Kaguru reckon time" as an aspect of their cosmology.[21] A similar concern with cosmology and conceptualization is apparent in more recent studies of the Tewa, the Navajo, and the Chamulas.[22] Susan Philips provides a fitting summary: "[P]eople of different cultures have different concepts of time."[23]

There is, then, ample evidence in support of the assertion that various societies have created unique and divergent forms of temporality. This assertion seems irreproachable, however, only at a rather distanced and abstract level of analysis. As it stands, the anthropological literature intellectualizes temporality by emphasizing formal and definitional matters, such as time-reckoning, cosmology, the meaning of time, temporal frames of reference, concepts of time, temporal systems or orientations, and different ways to speak of or refer to time—all of this to the neglect of variation in the perceived passage of time. One can observe an analogous tendency in the field of art criticism, where Susan Sontag has objected to an overemphasis on meaning and interpretation at the expense of direct experience.[24] Hence, it seems appropriate to paraphrase her rallying cry: In place of a hermeneutics, we need an erotics of temporality. At least it seems worth asking whether it is possible for there to be heterogeneity at one level of cross-cultural analysis (i.e., abstract conceptions of time) while there is homogeneity at another level of cross-cultural analysis (i.e., experiencing the passage of time in concrete circumstances).

A positive response to this question is implicit in the writings of Florence Kluckhohn:

All [societies] have some conception of the past, all have a present, and all give some kind of attention to the future time-dimension. They differ, however, in their emphasis on past, present, or future at a given period. . . .[25]

Her statement does not concern what I have called the elementary forms of temporality, but could it be that all societies also have some conception of variation in the perceived passage of time? A more ambiguous response can be found in *The Dance of Life*, by Edward Hall. He begins with the traditional argument: "[E]ach culture has its own time frames in which the patterns are unique."[26] Yet, in that same text, he seems to assert abstract principles that take the form of cross-cultural uniformities in experience: "The degree of concentration required to complete a task is related to how fast time is perceived as passing."[27] It is noteworthy that this quotation is not hedged about by the usual relativistic caveats. Indeed, its blunt avowal

suggests a level of analysis that transcends the idiosyncrasies of history and culture.

Hoyt Alverson asks the fundamental question: "Are
there universal as well as culturally particular experiences
and expressions of time?"[28] Based upon extensive linguistic
research, he contradicts Whorf by answering this question
in the affirmative:

In the course of working with two languages, English and
Setswana, during the three years I lived among the Tswana of
southern Africa, I observed that between two languages verbs or
nouns that designate what one might call mental states . . . in particular the notions of time—could in general be translated quite
exactly. . . . I have noticed this also in translating from English to
German, Mandarin, Hindi and other languages. . . . In my view,
despite significant discontinuities and differences, the interlan-
guage and intercultural translation of experience (like that of
time) is possible for two reasons: (1) we as humans share much
fundamental experience in common . . . irrespective of cultural
differences, and (2) the most important meaning-bearing aspects
of all human languages gear into and express that experience in
the same way.[29]

If, therefore, human beings must always and everywhere experience and express the perception that one thing is longer
(or shorter) than another, then there is reason to believe in
the possibility of a cross-cultural theory that accounts for
variation in the perceived passage of time.

Alverson argues that "anthropology and other human
sciences have overstated the . . . diversity of temporal experience."[30] In their efforts to describe temporal systems, an-
thropologists must, of necessity, render somewhat static

and stereotypical depictions.[31] What results is akin to the aforementioned claim by Durkheim "that a common time is agreed upon, which everybody conceives in the same fashion."[32] Correspondingly, there is less attention, if any, to intracultural variation and the dynamics of temporal experience—dynamics that may not be as divergent as the official system might lead one to believe. As previous chapters have shown, temporal experience in our own culture is often incommensurate with temporal frames of reference (or official time-reckoning). It seems reasonable to suspect that something like the same "deviance" is possible where temporal systems are not organized on the basis of clocks or calendars. Otherwise, we would have to assume that everyone in such cultures experiences the passage of time at the same rate. Moreover, it is not sufficient to leave the analysis at the level of linguistics, for we have seen how individuals often work around the impediments of language when trying to describe an uncanny temporal experience (uncanny precisely because one lacks the linguistic resources with which to articulate it easily).

In a famous aphorism, Clyde Kluckhohn and Henry Murray assert three levels of diversity in human experience:

EVERY MAN is in certain respects
   a. like all other men,
   b. like some other men,
   c. like no other man.[33]

The universal element (a) in their statement does not refer only to biological uniformity in our genus and species. As they are quick to point out, much of what is common in human experience is rooted in "societies and cultures"—an abstract phrase, to be sure, but one that presumably makes reference to the universal exigencies associated with socialization, social interaction, and the social construction of reality.[34] However, in his effort to define social interaction as a distinct topic of sociological research, it is Erving Goffman who has been most persuasive at sensitizing us to cross-cultural uniformities in conduct and experience:

Throughout this paper it has been implied that underneath their differences in culture, people everywhere are the same. If persons have a universal human nature, they themselves are not to be looked to for an explanation of it. One must look rather to the fact that societies everywhere, if they are to be societies, must mobilize their members as self-regulating participants in social encounters.[35]

Likewise, Norman Denzin "assumes that human affairs, wherever they occur, rest on the same interactional processes."[36] But what about temporality? Does it make sense to ask whether variation in the perceived passage of time exhibits cross-cultural uniformity? In response to questions of this sort, David Maines argues "that despite the great differences in cultural time logics and the substance of time-consciousness . . . various cultures have much in common when it comes to temporality."[37]

Obviously, these are encouraging statements, but they leave us with the following question: What circumstances could serve as the existential basis for a common pattern of variation in the perceived passage of time? We can begin to address this question, at least in speculative fashion, if we consider the following scenario from a short story by Italo Calvino:

I have the impression this isn't the first time I've found myself in this situation: with my bow just slackened in my outstretched left hand, my right hand drawn back, the arrow A suspended in midair at about a third of its trajectory, and, a bit farther on, also suspended in midair, and also at about a third of his trajectory, the lion L in the act of leaping upon me, jaws agape and claws extended.[38]

Let us assume that Calvino's narrator is not someone whose life is regulated by our standard temporal units. Is it not still reasonable to assume that, like various narrators in chapter 3, he will define this situation as one of extreme circumstances? Moreover, it seems equally reasonable to think that emotional concern and cognitive involvement will be heightened, thereby intensifying stimulus complexity and the density of experience, which, in turn, will bring about protracted duration. Indeed, the ensuing events may seem to transpire in slow motion.

If the foregoing scenario seems unrepresentative of daily life, then let us consider a more familiar problem. The folk saying, "A watched pot doesn't boil," refers to another situation in which there is the perception that time is passing slowly. In this case, however, there is too little happening instead of too much. But, once again, there is no need for recourse to clocks or calendars. Together, these scenarios imply that the people of divergent cultures can conceive of time in dissimilar ways even while experiencing its passage in much the same way. Is it possible, then, to specify a minimal model for the experience and expression of protracted duration? The prototypical statement could be phrased in the following fashion: "That day felt like a year." Only three conceptual elements are requisite, and they would seem to be universal, regardless of how divergent temporal systems might be otherwise. The individual in question would need a term for "day," a term for "year," and a term for "felt like." If future research can show that these conceptual elements are found in all cultural contexts (especially, of course, those that are quite different from our own), then there is reason to believe that people everywhere have the linguistic resources with which to experience and express the perception that events are transpiring slowly.

Again, however, we cannot leave the argument at the level of linguistics. Even if people in other cultures can be said to have the minimally necessary linguistic resources, is there any evidence that they make use of them? We can begin to address this question by looking briefly at data from an earlier period in our own society, since cultural evolution can make for significant differences between the temporal systems that characterize the same society at two different points in its history. Consider, for example, the great westward migration of settlers during the middle of the nineteenth century. Their journey was not organized on the basis of a contemporary concern with precise punctuality. But on Monday, September 15, 1862, Jane Gould made the following entry in her diary:

The road is the worst I ever saw. Lou and I walked the whole ten miles, till we came to within a mile of Ragtown. We saw the trees

on Carson River and thought we were almost there but we kept
going and going and it seemed as if I never could get there.[39]

Clearly, this woman is experiencing and expressing protracted duration, but neither the experience nor the expression relies on the standard temporal units of our clocks and
calendars.

For a more truly cross-cultural example, we can turn to
rural China—a part of the world where the rhythms of activity are governed by little more than the rising and setting
sun and the changing seasons. Yet, when a wedding sepa-
rates two sworn sisters, one writes to the other of her sor-
row in words that concern variation in the perceived passage of time:

Elder Sister being gone three whole days,
Feels like years.[40]

In chapter 3, we saw how suffering and intense emotions
(e.g., longing) can make for the experience of protracted
duration. Here, in a culture very different from our own, we
find the minimal model in pristine form.

Is the S-shaped model for variation in the perceived passage of time only applicable to our own culture? As we have
seen, there are eloquent arguments on both sides of this
issue. Ultimately, of course, arguments do not suffice, however eloquent they might be. This is, after all, an empirical
issue, and it will be resolved only as a result of systematic
observation. For now, there is just a glimmer of evidence,
but it seems to suggest the possibility of cross-cultural uniformity in temporal experience.

The Question of Human Agency

A theoretical model has emerged from my research, and it is capable of accounting for variation in the experience of duration. As it stands, however, the model is limited to analysis of how the situation and one's response to it condition the perceived passage of time. But what about the role of human agency in temporal experience? The relevance of this question becomes more apparent when we briefly turn our attention to neighboring fields of research: the social psychology of emotions and drug use.

Arlie Hochschild has demonstrated that emotional experience is not simply a response to the situation at hand, but is rather, at least in part, the result of intentional effort ("emotion work") on the part of the individual to promote or suppress emotional experience so as to be in accord with social expectations ("feeling rules").[41] Thus, as she puts it, "Sometimes we try to stir up a feeling we wish we had, and at other times we try to block or weaken a feeling we wish we did not have."[42] From her perspective, agency operates for the sake of "social order."[43] More recently, however, James Averill and Elma Nunley have argued that emotion work can (and should) be used for the sake of "emotional creativity" and self-actualization.[44] Although they emphasize the therapeutic effects of such agency, their argument implicates the purely recreational possibilities of emotion work. Of course, as Howard Becker notes, the same therapeutic and recreational alternatives confront those who are taking drugs:

We take medically prescribed drugs because we believe they will cure or control a disease from which we are suffering; the subjective effects they produce are either ignored or defined as noxious side effects. But some people take some drugs precisely because they want to experience these subjective effects; they take them, to put it colloquially, because they want to get "high."[45]

What we see in these studies is the human capacity to willfully modify subjective experience.

By analogy, I would anticipate that individuals engage in "time work" so as to promote or suppress a particular kind of *temporal* experience. To some extent, they probably do so for the sake of conformity with social expectations, but, drawing from the work of Averill and Nunley, I would also expect to find that individuals manipulate temporal experience for reasons of creativity and self-actualization. In his study of the art and culture of tattooing, Clinton Sanders describes the efforts of those who are engaged in "customizing the body."[46] Future research should reveal time work, agency, and self-actualization in the customizing of temporal experience. It follows that the question of human agency can be worded in the following fashion: To what extent and in what ways do individuals purposefully construct lines of activity or social situations in order to create a particular form of temporal experience?

There can be no doubt that variation in the perceived passage of time is governed by the nature of one's immediate circumstances. Instead of looking upon conduct and experience as products of one's situation, however, the foregoing question assumes a theoretical framework that can accommodate the analysis of agency or personal initiative in the construction of that situation to which one *wants* to respond (not because that situation is an end in itself, but because it is thought to bring about a particular kind of temporal experience). In what remains of this chapter, I want to build a theoretical foundation for the empirical study of time work, agency, and self-actualization in the customizing of temporal experience.

We must begin by conceiving of the individual as something more than a passive entity that merely reacts to environmental stimuli. If human beings can customize temporal experience, then they are capable of pausing in the face of competing stimuli and momentarily inhibiting their own responses while they consider and choose from among alternative lines of action. William James overstates the case for an active self, but his exaggeration serves to illuminate the often neglected role of choice in human experience:

Millions of items of the outward order are present to my senses which never properly enter into my experience. Why? Because they have no *interest* for me. *My experience is what I agree to attend to.*[47]

It would not be quite right to say that we "agree to attend to" an incipient collision between our car and a large animal that has ambled onto the road. Nonetheless, James does help us to understand how one can invoke what we might call the "savoring complex" by anticipating the desire to focus on a particular experience and then taking steps to ensure that distractions are kept to a minimum.

Advancing beyond the insight of James, George Herbert Mead recognized that this process of selective attention establishes a loop of mutual determination between the individual and his or her circumstances:

Our whole intelligent process seems to lie in the attention which is selective of certain types of stimuli. Other stimuli which are bombarding the system are in some fashion shunted off. We give our attention to one particular thing. Not only do we open the door to certain stimuli and close it to others, but our attention is an organizing process as well as a selective process. When giving attention to what we are going to do we are picking out the whole group of stimuli which represent successive activity. Our attention enables us to organize the field in which we are going to act. Here we have the organism as acting and determining its environment. It is not simply a set of passive senses played upon by the stimuli that come from without. The organism goes out and determines what it is going to respond to, and organizes that world.[48]

Mead creates a theoretical framework within which we can conceptualize self-determination by means of an interactionist construction of reality. From his standpoint, "action is now consciously purposive," and it involves "a process of selection from among various alternatives."[49] Put differently, he helps us to see how the individual exercises "choice" from among "alternative possibilities of future response."[50] One's circumstances, then, are not simply antecedent to one's response; rather, it is often the case that one plans for, and takes steps to bring into being, just those circumstances to which one would like to respond. Mead shows us that people can employ selective attention in an effort to organize a particular line of activity. I would add that, sometimes, they do so in order to evoke a special kind of temporal experience.

As different as they are in other respects, the writings of George Herbert Mead and Erving Goffman dovetail quite nicely on the topic of agency.[51] What is more, Goffman's earliest observations set the stage for Hochschild's conceptualization of emotion work: "[E]ach participant is expected to suppress his immediate heartfelt feelings, conveying a view of the situation which he feels the others will be able to find at least temporarily acceptable."[52] In this statement, Goffman demonstrates his typical concern with outward display, while Hochschild's formulation is primarily concerned with the modification of inner experience. Laying aside this difference in emphasis, however, we should recognize that Goffman makes an important contribution to the construction of a theoretical paradigm that renders intelligible one's efforts to manipulate or control the circumstances to which one must respond:

Whether a social situation goes smoothly or whether expressions occur that are in discord with a participant's sense of who and what he is, we might expect—according to the usual deterministic implications of role analysis—that he will fatalistically accept the information that becomes available concerning him. Yet when we get close to the moment-to-moment conduct of the individual we find that he does not remain passive in the face of the potential meanings that are generated regarding him, but, so far as he can, actively participates in sustaining a definition of the situation that is stable and consistent with his image of himself.[53]

Like Hochschild, Goffman attributes these efforts to motives that are self-serving (i.e., concerned with avoiding trouble) but generally salutary for the maintenance of social order in face-to-face settings. He does little to acknowledge that someone might manipulate his or her circumstances for reasons of self-actualization (i.e., for the sheer fun of it or for aesthetic purposes). For that component of the theoretical framework, we must turn to the writings of other scholars.

Abraham Maslow put self-actualization at the top of his hierarchy of "basic needs."[54] He argues that we have "been ducking the problem of . . . will" by overemphasizing the determinism of the past (X causes Y, etc.) and overlooking the indeterminism of agency, which involves our deliberate attempts to shape the future.[55] Predating Averill and Nunley, Maslow calls for a pleasure-driven theory of motivation that embraces one's "enjoyable" and "creative" efforts to be the architect of one's own circumstances.[56] His concerns are echoed by contemporary scholars. Mihaly Csikszentmihalyi accuses the social sciences of neglecting "the dynamics of intrinsic motivation" (i.e., "activities that appear to contain rewards within themselves").[57] Likewise, Jack Katz is critical of criminology for not facing up to the fact "that the causes of crime are constructed by the offenders themselves."[58] Indeed, he demonstrates that, rather than being pushed into crime by the background factors which typify the disadvantaged, many of those who pursue criminal activity choose to do so because they are drawn to "the positive, often wonderful attractions within the lived experience of criminality."[59]

Summarizing recent trends in social psychology, Elizabeth Menaghan asserts that "the individual is increasingly conceived as an active agent who may be more powerful in shaping his or her own trajectory and even in altering social arrangements than prior formulations have recognized."[60] Thus, there is a theoretical framework that points to the possibility of time work, agency, and self-actualization in the customizing of temporal experience. The theoretical model that has emerged from this study lays the necessary foundation for the investigation of the folk theories and methods employed by those who would manipulate their circumstances in an effort to bring about a particular form of temporal experience. It only remains to map the empirical contours of these efforts, and preliminary evidence gives us an idea of what to look for.

In *Rivethead: Tales from the Assembly Line*, Ben Hamper describes his "favorite method" for speeding up the perceived passage of time:

What I would do was to pretend my job was an Olympic event. I would become both television narrator and participant. It would go something like this:

"We've come to the end of another long day for the American squad. So far, the Japanese have totally dominated each event, sweeping the gold and silver in every category. Any final hopes the Americans might have of winning a gold medal rest solely on the shoulders of Ben Hamper, an assembler out of the GM Truck and Bus facility in Flint, Michigan.

"Hamper will be competing in the Freestyle Rivet Squash. Though he's considered a long shot at best, you will see in this recent interview that Hamper believes he's capable of pulling off the upset."[61]

Hamper goes on to describe his interview with Jim McKay, his frantic efforts to set a new world record, and the subsequent celebration. At the end of this virtuosic fantasy, Hamper has won a very real victory over time: "The cheering faded away and, with it, a few more minutes off the clock."[62]

Some wag once noted that time is nature's way of keeping everything from happening at once. Certainly this is true of research. In this study, some steps have been taken toward understanding variation in the perceived passage of time. But upon cresting each hill, we find another vista of uncharted terrain. There is, then, important work to be done, and it holds the promise of expanding the horizon of our knowledge in social psychology.

Methodological Appendix

A methodological appendix provides an opportunity to pause and reflect on the conduct of one's own research. So, in the following remarks, I want to discuss some principles that have guided my investigation of variation in the perceived passage of time.[1] Interestingly enough, over the course of my research, I have thought a lot more about method than I have about truth or validity. Yet, as I look back on it, issues pertaining to truth and validity have haunted what seemed, at the time, my purely technical concerns with the mechanics of method.

This appendix is in keeping with "the reflexive turn taken by ethnographic practice, beginning 20–25 years ago."[2] Indeed, John Johnson and David Altheide argue that methodological reflexivity is the distinguishing attribute of social science texts.[3] Its contemporary prominence is related to a number of issues pertaining to the author's place in the text, and different styles arise as authors address these issues in divergent ways.[4] Thus, George Marcus identifies four styles of methodological reflexivity: subjectivist, objectivist, intertextual, and feminist. With its "commitment to sustain objectivity, the distance and abstraction of theoretical discourse, and empiricism as distinctive historical contributions of sociology," the objectivist style is most akin to my own.[5] This kinship will become apparent as I summarize the pursuit of my substantive quarry: a theoretical model that accounts for the full range of variation in the perceived passage of time.

Methodological reflexivity is represented by a systematic account of the methods used in one's own research. As such, it should not be confused with methodology, which is the study of methods in the abstract. Of course, this is an analytical distinction, and, in actual practice, the authors of textbooks and chapters in edited collections frequently draw examples from their own work.[6] Typically, however, they do so in rather haphazard fashion, and, along the way, we learn a good deal more about methods in the abstract than we learn about their own particular methods. It is rare, then, to find an exercise in methodological reflexivity standing alone, and this is to be expected, given the necessary emphasis on discovery rather than reflection.[7] Nonetheless, there is something of value in methodological reflexivity, especially during an era marked by widespread uncertainty concerning the practice of social research.

In this appendix, I reflect on the conduct of my own research and discuss five principles that are pertinent to validity, truth, and method in the study of temporal experience. These principles are as follows: intellectual curiosity, systematic empiricism, analytic induction, a pragmatic orientation, and a commitment to progress. I conclude by calling for a revitalization of these principles in social research.

Intellectual Curiosity

I have been studying variation in the perceived passage of time for roughly ten years, but my memory of how it all got started is still quite fresh. I am sitting with other students in a graduate seminar when Norman Denzin strolls into the room with a document he finds intriguing. It is an interview with John Brodie, a former quarterback with the San Francisco 49ers, and he is talking about how aspects of his perception change during the course of a game:

At times, and with increasing frequency now, I experience a kind of clarity that I've never seen adequately described in a football story. Sometimes, for example, time seems to slow way down, in an uncanny way, as if everyone were moving in slow motion. It seems as if I have all of the time in the world to watch the receivers run their patterns, and yet I know the defensive line is coming at me just as fast as ever.[8]

Today, I cannot recall exactly how the interview was relevant to our seminar, but I do remember being particularly struck by this quotation. There are any number of reasons why I find this statement fascinating, but it is fundamental to my own epistemological stance that the reasons are not terribly interesting or, for that matter, even germane. In any event, I have no idea what they might be. And what difference does that make? Motivational considerations are tangential if, like me, you assume that the audience is more interested in your findings and methods than your reasons for starting the research in the first place. So, once again, I am relating how I have studied something, and not why I have studied it.

Howard Becker decries the sociological tendency to "mak[e] common events and experiences mysterious."[9] In particular, he is critical of elaborate theories that purport to account for deviant motivation. Likewise, motivation for research is mystified when it is attributed to the psychological or sociological background of the researcher. Either way, one is presumed to act on the basis of forces beyond one's ken. This is not in keeping with a symbolic interactionist understanding of human conduct, and it does not square with the motivation for my own research. Norman had presented us with a puzzle to be solved; I provided the inquisitiveness. Perhaps there is some irony in the fact that research, like deviance, often begins with nothing more than a combination of curiosity and opportunity.[10]

Thus, the first principle is that intellectual curiosity is sufficient to provoke research. Any further consideration of motivation is irrelevant to an effort at methodological reflexivity. But how do you proceed once you are curious about something? The second and third principles are the twin pillars of the social sciences: dedication to systematic empiricism and concentration on the analysis of process and meaning.

Systematic Empiricism

I began to watch for statements that resembled Brodie's, but, for a long while, this activity was subordinate to my usual readings of the popular press. Nonetheless, having become sensitized to something that had always been there, I began to find and record other examples. I did not take this activity very seriously at first. It was interesting, but it resembled a hobby, like stamp collecting, more than my other research projects, and I did not have big plans for the accumulating documents. This was especially so because, despite my interest, gathering "found" data is a slow process and there was very little of it for quite some time. Still, I kept watching for other examples, and, at last count, there were 389 separate records. Moreover, the data set continues to grow.

Along the way, I started to take the project more seriously because I could see that it had exciting possibilities. My interest grew in proportion to the data gathered. Indeed, at some point, what had been a subordinate activity became my principal project. I began to recognize recurrent themes, and I speculated about their origins. In addition, it was apparent that the topic might have important implications for both theory and research in social psychology. My enthusiasm was tempered by the fact that the data were somewhat foreign to one trained in the analysis of ethnographic field notes. However, we had been reading *Frame Analysis* in Denzin's seminar, and Goffman had set something of a precedent by having made impressive use of stories taken from the popular press.

I crossed a significant threshold when I began to anticipate finding data in certain documents. I could look at the headline of an article or the title of an autobiography and know that new evidence could be found there. I took this as affirmation of my growing familiarity with its experiential niche.[11] On the one hand, the pace of the data gathering quickened as I learned where to look for observations. On the other hand, serendipitous findings continued to broaden the range of my inquiry. I conceived of my quarry as situations in which people experienced time passing slowly. The obvious candidates involved waiting or boredom, but, as the interview with Brodie indicated, the experience is also occasioned by situations that can only be defined as exciting or challenging. And, in fact, it seemed reasonable to think that the latter are more likely to appear in the popular press. Consequently, the data were accumulating, but personal reflection suggested that my method might not be capturing the full extent of variation. Furthermore, it is in the nature of "found" data that one takes only what one finds, as there are no opportunities to probe someone's responses for the sake of elaboration.[12]

Despite these shortcomings, the use of "found" data had much to recommend it. For one thing, this technique enabled me to obtain empirical materials from a wide variety of cultural and historical contexts. What is more, while there is a certain self-consciousness provoked by being quoted for attribution or writing one's memoirs, the character of that "reactivity" differs from that of people (tellingly called "subjects") who know they are participating in a research project.[13] Finally, this procedure is capable of producing rich data, especially when the stories are told by articulate narrators describing distortion in the perceived passage of time, such as Aldous Huxley under the influence of mescaline, Albert Camus ruminating on the effects of imprisonment, and Joan Didion driving on the Santa Monica Freeway.[14] In any event, with repetition increasing and serendipity decreasing, I could see the onset of "theoretical saturation."[15]

For a variety of reasons, then, the emphasis of my research shifted from the relatively passive approach of simply finding data, to the relatively active one of generating data through in-depth interviews. This shift in emphasis allowed for the "triangulation" of methods with different advantages and disadvantages.[16] There was also a sense in which I viewed this transition as an in-house effort at replication. In other words, I wanted to see if there would be confirmation of the findings derived from my previous data. Moreover, asking questions of cooperative subjects enabled me to probe interesting answers for elaboration. Generally speaking, the resulting transcripts dovetailed nicely with the earlier data. However, there were fewer incidents of violence (e.g., earthquakes, automobile accidents, and shootouts) and more examples of suffering (e.g., dentistry, wind sprints, and painful ailments). These differences provided some balance for the bias toward excitement in the popular press. And, of course, interviewing produced empirical materials more quickly than did my earlier method because I was generating data rather than finding it.

Between 1986 and 1988, my assistants and I conducted interviews with 316 undergraduates enrolled in several sections of an introductory sociology course. It was our intention to elicit stories that were comparable to the narratives I had found in the popular press. Our script for these interviews did not specify the exact wording of each question, but we did use a schedule to guide ourselves through the same sequence of questions. The interviews were retrospective in that we would ask respondents to recollect an autobiographical episode in which time had been perceived to pass slowly. A general prompt of this sort helped to bring the details of an incident to the surface of consciousness.

After confirming the appropriateness of the story, we would proceed with a series of questions that required respondents to write a detailed summary of the episode, including their perception of the passage of time.

We asked our respondents to recall the day, date, and time of the incident (to the best of their ability). Then we asked them to describe the physical setting, relevant objects, the nature of the social situation, and the people who were involved in that episode. Finally, we asked our respondents to retell their stories while being careful to relate the sequence of events and how those events affected their perception of the passage of time. In effect, we prompted our respondents to reconstruct the elements of a situation in which they had experienced time passing slowly. The format of the interview was designed to produce narratives such that it was as if the researcher had *been* the respondent, present at the scene, and thereby able to record not only the objective features of the situation, but also his or her own subjective experiences of temporality.[17] Our use of interviewing as a data collection strategy ended only when it became clear that, once again, we had achieved theoretical saturation.[18]

The triangulation of data and methods is a crucial aspect of systematic empiricism. Eugene Webb and his associates advocate "multiple operationism," and they point out the fundamental advantage that accrues from the use of multiple methods: "The most persuasive evidence comes through a triangulation of measurement processes."[19] Webb and his associates seem to have one's audience in mind, but my own feeling of certainty grew as it became obvious that, for the most part, the interviews were replicating the found data.

By way of summary, then, the second principle entails dedication to systematic empiricism. The resulting observations should be eclectic as well as exhaustive. There should be evidence that one has made at least a good faith effort to describe the full range of variation. But most critically, it should be apparent that the researcher has given the all too recalcitrant world of empirical materials an honest chance to say "no, you're wrong."

Analytic Induction

There is, then, the possibility of being wrong, and this possibility implicates the third methodological principle: one's analysis should reveal universal aspects of process and meaning.[20] In *The Principles of Psychology*, William James states that "*a tract of time empty of experiences seems long in passing.*"[21] Try telling that to John Brodie or, for that matter, the narrators of roughly half of my 705 cases. To paraphrase Goffman, the statement by James is right as it reads but wrong as it is taken.[22] In fact, the statement is applicable only to a very narrow segment of the data having to do with waiting and boredom. James would have known this had he been dedicated to systematic empiricism. Thus, from a broader standpoint James is wrong, at least insofar as he intends his statement to embrace the full range of variation, and serve as a general explanation for the phenomenon in question.

A single criterion guided my analysis of process and meaning: if what you say is only partially correct, and is, moreover, contradicted by what you neglect to say, then some people will view your statement as misleading while others will view it as downright wrong. This criterion is derived from the tradition of analytic induction, a tradition that, in sociology, can be traced to the writings of Florian Znaniecki.[23] As Alfred Lindesmith puts it, "[T]he cause of a phenomenon is that complex of conditions, or that process, without which the phenomenon cannot occur and in the presence of which it never fails to occur."[24] Howard Becker adds that a commitment to analytic induction involves repeatedly modifying one's working hypotheses throughout the course of one's research:

The method requires that *every* case collected . . . substantiate the hypothesis. If one case is encountered which does not substantiate it, the researcher is required to change the hypothesis to fit the case which has proven his original idea wrong.[25]

Bear in mind that the "original idea" is "wrong" not because it is an outright falsehood, but because it fails to encompass the full range of variation. Peter Manning provides a helpful summary: "Analytic induction seeks to develop *universal* statements containing the *essential features* of a phenomenon."[26]

In the beginning, my analysis centered on the following question: What circumstances provoke the perception that time is passing slowly? The answer coalesced from a variety of sources: my personal experiences, my reading of previous research on this topic, and my analysis of a set of empirical materials that gradually grew larger and more diverse. In hindsight, there were other reasonable questions on which to focus the project, but the perception that time is passing slowly seemed to be the best place to start. It was, after all, Brodie's experience that had sparked my interest to begin with. Furthermore, as a response to unusual circumstances, it is likely to appear in the popular press and people are prepared to discuss such incidents during an interview.

In keeping with the goal of analytic induction, I tried to formulate concepts, hypotheses, and theory that would circumscribe the complete range of variation in my empirical materials. My analysis was inductive and oriented toward theory construction. From this standpoint, thorough description was necessary but not sufficient. Put differently, description was not an end in itself, but merely a means to an end. I was trying to develop a general explanation—one that would account for all of my observations. Consequently, my concepts, working hypotheses, and theoretical framework were revised repeatedly in response to the appearance of what Lindesmith calls "negative cases."[27]

The use of analytic induction required that I sift out concrete but irrelevant details in order to reveal common and therefore essential factors. Presumably, these common factors were present whenever time was perceived to be passing slowly. This approach entailed a process of abstraction as, step by step, I moved from the idiosyncrasies of this or that situation to the general properties of theoretical explanation. However, what did a football game have in common with imprisonment, the use of hallucinogenic drugs, and a boring conversation? These and other disparate situations produced the same temporal experience, but why? When I finally arrived at an interpretation that fit the facts, it no longer resembled the narrative materials with which I had started. Through successive stages, my analysis had evolved into the interlocking propositions of formal theory.

The logic of analytic induction leads the investigator from observations to generalizations. What is more, Denzin notes that analytic induction has been the predominant approach to theory construction among students of social interaction:

Formal theory . . . is a common goal of interactionist research. Although historically, or situationally, specific propositions are recognized, propositions with the greatest universal relevance are sought.[28]

Through the use of analytic induction, I have tried to formulate a universally valid understanding of those circumstances that govern variation in the perceived passage of time.

Analytic induction is a form of social interaction. As such, it is characterized by process and reflexivity. The use of analytic induction requires constant comparison of observations with interpretations.[29] A related process occurs in ordinary interaction; one continually evaluates the appropriateness of one's responses to the gestures and utterances of those with whom one interacts. Do the accumulating observations fit the emerging theory? Negative cases, like untoward gestures, demand reformulation of the researcher's always tentative understanding. Of course, this perspective is predicated upon Herbert Blumer's conception of research as a form of interaction with the world:

> [T]he empirical world can "talk back" to our pictures of it or assertions about it—talk back in the sense of challenging and resisting, or not bending to, our images or conceptions of it. This resistance gives the empirical world an obdurate character that is the mark of reality.[30]

It is not uncommon to misinterpret the gestures of others, and, within the context of research, a finite number of observations cannot ensure that a particular theory is correct. However, observations do enable us to recognize when a particular theory is wrong.

A Pragmatic Orientation

My qualitative data and analytic induction had produced a theoretical model which, on the face of it, accounted for the experience of protracted duration and synchronicity. But what about temporal compression? As yet, the theoretical model contained no explanation for the perception that time has passed quickly. The evidence at hand indicated that people experience time passing slowly when their subjective involvement with immediate circumstances is high. And, as a taken-for-granted aspect of background expectancies, time passes in largely unnoticed fashion under ordinary circumstances where subjective involvement is moderate. By a process of elimination, this left a situation characterized by low subjective involvement as the context for temporal compression. But under what circumstances do people act without paying much attention to what they are doing? Moreover, why are they not bored (thereby experiencing protracted duration)?

When I put this question to one of my colleagues, Jim MacDougall, he said that I was describing what cognitive psychologists call automatic processing—in other words, conduct so habitual that one need not pay much attention to it. This seemed to be a very promising lead, and I grew still more enthusiastic when Ralph Turner and Victoria Billings reminded me that the founding fathers of symbolic interactionism, especially Mead and Dewey, were intent on understanding the difference between habitual and problematic circumstances.[31] Unwittingly, my efforts to conceptualize temporal compression had led me back to core concerns of social psychology. Still, I was not sure how empirical materials could be used to substantiate the suspected relationship between habitual conduct and the perception that time has passed quickly. This problem was derived from greater ambitions for the project than those with which I had started, but each door seemed to open on yet another door.

I needed a research design that would allow me to study the temporal experience of people operating within two different contexts: one characterized by routine complexity, the other by problematic complexity. My theory predicted that temporal compression would result from the former and protracted duration from the latter. I was considering various possibilities when correspondence from Gary Fine rendered the issue moot. He asked me to read a manuscript that would later become his article, "Organizational Time: Temporal Demands and the Experience of Work in Restaurant Kitchens." His study made it clear that chefs face two kinds of "rush": there is the rush for which they are prepared, resulting in routine work that seems to have gone by quickly; and there is the rush for which they are unprepared, resulting in problematic work that seems to go by slowly. Gary's findings confirmed what my theory would predict concerning the relationship between habitual conduct and temporal compression. Moreover, his findings helped me to realize that there are two kinds of busy time: routine and problematic. This insight allowed me to reconcile my data on protracted duration with the folk saying, "Busy time passes quickly," thereby integrating the conceptualization of temporal compression with the rest of my theoretical model. In addition, Gary introduced me to the writings of Csikszentmihalyi, which dovetailed nicely with Gary's findings and provided further corroboration for my own hypothesis.[32]

Nonetheless, I could see that further conceptualization was called for. Participation in routine complexity is intermittent and only characteristic of particular situations, but everyone seems to agree that, generally speaking, time has passed quickly. A variable cannot be used to account for a constant. Clearly, something was missing—something correlated with (and therefore just as prevalent as) the feeling that time has passed quickly. After giving this problem some thought, it finally occurred to me that loss of memory was the missing piece of the puzzle. Consequently, I began to read research on memory by cognitive psychologists (again, at the instigation of Jim MacDougall).[33] These studies did not address temporality, but what I learned from them lent credence to the working hypothesis derived from my theoretical model. The model predicts that one will experience temporal compression when conscious information processing is low. Given the ubiquitous loss of episodic memory over time, it is logical to assume that all but a few remembered intervals are contracting as time passes, thereby seeming to have passed much faster than they actually did.

Once again, the theoretical model had been extended in plausible fashion, but, by itself, face validity seemed inadequate. In other words, I felt a need for the kind of confirmation that could only result from examining the hypothetical relationship between memory and temporal compression in light of new empirical materials. But how? The empirical implications concerned erosion in the volume or *amount* of remembered experience per standard temporal unit. Clearly, deductive logic and quantitative methods were called for, but how could we measure the erosion of episodic memory and show that such erosion is associated with temporal compression? By posing this question, I assumed that the credibility of our research is enhanced when we "test our hypotheses and propositions against empirical evidence."[34]

My first plan involved asking subjects to write narratives about their activities and experiences yesterday, last month, and last year. Given the erosion of episodic memory, I hypothesized that the subjects would write longer stories about yesterday than they would about last month, and longer stories about last month than they would about last year. The preliminary results from our pretests were quite encouraging, but it quickly became obvious that this procedure would require more cooperation than many of the subjects were willing to give. In addition, it was a very time-consuming process and not at all conducive to the rapid collection of data from large groups of people. Consequently, we decided to have our subjects tell us how they perceived time to have passed yesterday, last month, and last year by selecting a number from each of three Likert-type scales where 1 = Very Slowly, 2 = Slowly, 3 = Normally, 4 = Quickly, and 5 = Very Quickly.

The foregoing summary suggests a pragmatic orientation toward research, and I submit that we make this our fourth principle.[35] Every form of inquiry is characterized by certain disadvantages, so it is inevitable that objections will be raised concerning one's methods. However, such objections are only constructive so long as they provoke further research. If the objections dampen enthusiasm for research, or, worse, question the possibility of ever getting at the truth of the matter, then the objections are antithetical to the epistemology of the social sciences.

A Commitment to Progress

Like a castle made of sand, every theoretical model is provisional; eventually it is washed away by the tide of continued research. Either new observations arise which do not fit within the theory's explanatory framework, or new conceptual developments emerge which enable us to construct a more parsimonious model. Put differently, new research continuously challenges the efficacy of extant theories. In so doing, it allows us to choose from among competing models by confronting each of them with at least two questions: Does that theory encompass the full range of variation in the empirical materials? And does that theory provide a logically coherent explanation for that variation? Throughout this project, my goal has been a theoretical model that answers both of these questions in the affirmative—a model, moreover, that advances our understanding of lived duration. As such, this whole endeavor is predicated upon the perfectibility of formal theory.

Progress may be the ultimate goal, but the world is a large and inconstant place, and the prevalence of social change can make this goal seem elusive if not illusory. Admittedly, there are always transient aspects to the contemporary scene—in other words, forms of conduct and experience that are creatures of historically specific conditions. But there are also perduring forms of conduct and experience—phenomena that are associated with unchanging aspects of self, interaction, and the social construction of reality. In his essay, "The Dual Mandate of Social Science," Everett Hughes refers to this distinction with the terms "time-bound" and "timeless."[36] My own work suggests an ironic juxtaposition—a theory concerning variation in the perceived passage of time which reveals timeless properties of temporal experience. But, of course, only further research can confirm the validity of this statement. There is, then, the possibility of "discovery" in what George Marcus calls "the classic sense" of that word—discovery, moreover, that makes for progress in our understanding of human nature.[37]

Howard Becker has written that "given a question and a method of reaching an answer, any scientist, whatever his political or other values, should arrive at much the same answer, an answer given by the world of recalcitrant fact that is 'out there.'"[38] I embrace the "realist" assumptions implicit in that passage.[39] Typically, however, the issue is not one of objectivity (i.e., do we have "the" answer?), but one of intersubjectivity (i.e., do we have evidence and argument that colleagues find persuasive?). Thus, we must recognize that, from a practical standpoint, truth and validity are intersubjective concepts. This is not an argument for self-indulgence or autistic solipsism. Rather, it is acknowledgment of what Mead might have called a methodological "generalized other"—that is, an understanding (imperfect though it may be) of what colleagues will ratify as appropriate methods.[40] In other words, truth and validity are what pass for truth and validity within a particular "thought collective."[41]

Recognizing the social context of inquiry, indeed, of all knowledge, can lead to cynicism and resignation. In recent years, we have witnessed an upsurge of doubt and genuine confusion as well as skepticism and what I would characterize as an opposition to inquiry.[42] It may well be that in the rough and tumble world of empirical research, we only ever have "approximations to knowledge."[43] But I would argue that, with continued effort, these approximations grow ever more proximate to patterns of conduct and structures of experience that are more than mere artifacts of method, rhetoric, and "prevailing thought style."[44] The fifth principle is, therefore, an optimistic belief in the possibility of progress and the perfectibility of theoretical understanding. Symbolic interactionism is capable of scientific, that is to say, verifiable generalization, explanation, and prediction. In contrast to so much of what is fashionable, what I am describing is an American epistemology, and one that is in keeping with the roots of symbolic interactionism.[45]

Conclusion

Leaving our offices and libraries for the field is often a difficult, time-consuming process, and one fraught with any number of good reasons for doubting its efficacy. Perhaps that is why we spend far too much time trying to figure out what classical and contemporary theorists said and did, and not nearly enough time finding out what nameless people on the street say and do. In any event, the real and imagined problems of method should not be allowed to discourage our investigation of social interaction.

I would like to think that these comments could be taken for granted, but my experience as a reviewer and editor does not suggest that this is the case. So, today, as I reflect on the study of lived time, I see the need for revitalization of our commitment to research. More specifically, the research which I envision is born of intellectual curiosity, sustained by systematic empiricism, and directed toward the discovery of what Carl Couch has called "abstract sociological principles."[46] This kind of inquiry will broaden and deepen our understanding of the relationship between self and society.

My own research has been inductive and deductive, descriptive and explanatory, qualitative and quantitative. I offer no apologies for all of this expediency, and happily cede consistency to those with programmatic positions to defend. For my part, I have pursued elusive quarry across challenging terrain, always doing whatever seemed necessary to keep the quarry in sight.

Notes

1. William James, *The Principles of Psychology*, Vol. 1 (New York: Henry Holt, 1890).

2. Thus, time seems to pass slowly in an environment that changes only incrementally:

> Time, Armstrong would later note, was a strange commodity on the moon. While their mission proceeded with an accuracy of minutes or seconds, he and Aldrin were on a world where a day lasts a month, where time seems to crawl. Looking at this landscape of craters, rocks, and dust he had the feeling that he was seeing a snapshot of a world in steady-state, that if he had been here a hundred thousand years ago or if he returned a million years from now he would see basically the same scene.

See Andrew Chaikin, *A Man on the Moon: The Voyages of the Apollo Astronauts* (New York: Viking, 1994), pp. 217–18.

3. Edmund Husserl, *The Phenomenology of Internal Time-Consciousness*, ed. Martin Heidegger, trans. J. S. Churchill (Bloomington, Ind.: Indiana University Press, [1928] 1964), p. 60.

4. Henri Bergson, *Duration and Simultaneity* (Indianapolis, Ind.: Bobbs-Merrill, [1922] 1965), p. 44.

5. Martin Heidegger, *Being and Time*, trans. J. Macquarrie and E. Robinson (London: SCM Press, [1927] 1962).

6. George Herbert Mead, *Mind, Self, and Society* (Chicago: University of Chicago Press, 1934).

7. Emile Durkheim, *The Elementary Forms of the Religious Life*, trans. J. W. Swain (Glencoe, Ill.: Free Press, [1912] 1954), p. 492.

8. David S. Landes, *Revolution in Time: Clocks and the Making*

of the Modern World (Cambridge, Mass.: Belknap/Harvard, 1983); Lewis Mumford, *Technics and Civilization* (New York: Harcourt, Brace, 1934).

9. Georg Simmel, "The Metropolis and Mental Life," in *Georg Simmel: On Individuality and Social Forms*, ed. N. Levine (Chicago: University of Chicago Press, [1903] 1971), p. 328.

10. William James, *The Principles of Psychology*, Vol. 2 (New York: Henry Holt, 1890), p. 291.

11. Alfred Schutz, *Collected Papers*, Vol. 1, *The Problem of Social Reality*, ed. M. Natanson (The Hague: Martinus Nijhoff, 1962), p. 207.

12. Ibid., p. 230.

13. Erving Goffman, *Frame Analysis: An Essay on the Organization of Experience* (Cambridge: Harvard University Press, 1974).

14. Georges Gurvitch, *The Spectrum of Social Time*, trans. M. Korenbaum (Dordrecht, Holland: D. Reidel, 1964), p. 20; Edward T. Hall, *The Dance of Life: The Other Dimension of Time* (Garden City, N.Y.: Anchor Doubleday, 1983), p. 139.

15. Alfred Schutz and Thomas Luckmann, *The Structures of the Life-World*, trans. R. M. Zaner and H. T. Engelhardt, Jr. (Evanston, Ill.: Northwestern University Press, 1973), p. 56.

16. Husserl, *The Phenomenology of Internal Time-Consciousness*, p. 22.

17. James Joyce, *Ulysses* (New York: Modern Library, [1922] 1961).

18. Mead, *Mind, Self, and Society*.

19. Anselm L. Strauss, "Mead's Multiple Conceptions of Time and Evolution: Their Contexts and Their Consequences for Theory," *International Sociology* 6 (1991): 411–26.

20. George Herbert Mead, *The Philosophy of the Present* (Chicago: University of Chicago Press, 1932), p. 31.

21. George Herbert Mead, *The Philosophy of the Act* (Chicago: University of Chicago Press, 1938), p. 65.

22. Mead, *Mind, Self, and Society*, p. 176.

23. David R. Maines, Noreen M. Sugrue, and Michael A. Ka-

tovich, "The Sociological Import of G. H. Mead's Theory of the Past," *American Sociological Review* 48 (1983): 161–73.

24. Mead, *The Philosophy of the Act*, p. 66.

25. Ibid., p. 151; see also Norman K. Denzin, "Act, Language, and Self in Symbolic Interactionist Thought," *Studies in Symbolic Interaction* 9 (1988): 53.

26. Mead, *The Philosophy of the Act*, p. 160.

27. Ibid., p. 54.

28. Denzin, "Act, Language, and Self in Symbolic Interactionist Thought," pp. 60–61, italics in original.

29. Denzin provides some justification for the change in emphasis: "Consciousness is the factual starting point of an interpretive theory of experience in the world." Ibid., p. 75.

30. Mead, *The Philosophy of the Present*, p. 13.

31. Mead's own reservations notwithstanding, I concur with Denzin's argument that "the main tenets of symbolic interactionist thought . . . are compatible with . . . phenomenology." See Mead, *The Philosophy of the Act*, p. 35; Norman K. Denzin, "Emotion as Lived Experience," *Symbolic Interaction* 8 (1985): 224.

32. Heidegger, *Being and Time*; Alfred R. Lindesmith, Anselm L. Strauss, and Norman K. Denzin, *Social Psychology*, 7th ed. (Englewood Cliffs, N.J.: Prentice Hall, 1991), p. 7.

33. Heidegger, *Being and Time*, p. 39. Schutz elaborates on this idea when he argues that each "finite province of meaning" is characterized by "a specific time-perspective." See *Collected Papers*, Vol. 1, p. 230.

34. Norman K. Denzin, "On Time and Mind," *Studies in Symbolic Interaction* 4 (1982): 38.

35. Husserl, *The Phenomenology of Internal Time-Consciousness*, p. 79.

36. Ibid., p. 52.

37. Ibid., p. 47.

38. Eugène Minkowski, *Lived Time: Phenomenological and Psychopathological Studies*, trans. N. Metzel (Evanston, Ill.: Northwestern University Press, [1933] 1970), p. 15.

39. Ibid., p. 25.

40. Ibid., p. 83. See also p. 161.

41. Ibid., p. 18.

42. Pitirim A. Sorokin and Robert K. Merton, "Social Time: A Methodological and Functional Analysis," *American Journal of Sociology* 42 (1937): 615–29.

43. Gary Alan Fine, "Organizational Time: Temporal Demands and the Experience of Work in Restaurant Kitchens," *Social Forces* 69 (1990): 95–114; David R. Maines and Monica J. Hardesty, "Temporality and Gender: Young Adults' Career and Family Plans," *Social Forces* 66 (1987): 102–20; Murray Melbin, "Behavior Rhythms in Mental Hospitals," *American Journal of Sociology* 74 (1969): 650–65; idem, "Night as Frontier," *American Sociological Review* 43 (1978): 3–22; Robert K. Merton, "Socially Expected Durations: A Case Study of Concept Formation in Sociology," in *Conflict and Consensus*, ed. W. W. Powell and R. Robbins (New York: Free Press, 1984); Wilbert E. Moore, *Man, Time, and Society* (New York: John Wiley and Sons, 1963).

44. Eviatar Zerubavel, "Timetables and Scheduling: On the Social Organization of Time," *Sociological Inquiry* 46 (1976): 87–94; idem, "The French Republican Calendar: A Case Study in the Sociology of Time," *American Sociological Review* 42 (1977): 868–77; idem, *Patterns of Time in Hospital Life* (Chicago: University of Chicago Press, 1979); idem, *Hidden Rhythms: Schedules and Calendars in Social Life* (Chicago: University of Chicago Press, 1981); idem, "Easter and Passover: On Calendars and Group Identity," *American Sociological Review* 47 (1982): 284–89; idem, "The Standardization of Time: A Sociohistorical Perspective," *American Journal of Sociology* 88 (1982): 1–23; idem, *The Seven Day Circle* (New York: Free Press, 1985).

45. Eviatar Zerubavel, *The Fine Line: Making Distinctions in Everyday Life* (New York: Free Press, 1991), p. 6. See also Eviatar Zerubavel, *Social Mindscapes: An Invitation to Cognitive Sociology* (Cambridge: Harvard University Press, 1997).

46. Fred Davis, "Definitions of Time and Recovery in Paralytic Polio Convalescence," *American Journal of Sociology* 61 (1956): 583; Julius A. Roth, *Timetables: Structuring the Passage of Time in Hospital Treatment and Other Careers* (Indianapolis, Ind.: Bobbs-Merrill, 1963), p. 12.

47. Barney G. Glaser and Anselm L. Strauss, *Time for Dying* (Chicago: Aldine, 1968), pp. 5 and 14.

48. David R. Maines, "Time and Biography in Diabetic Experience," *Mid-American Review of Sociology* 8 (1983): 109; Helena Z. Lopata, "Time in Anticipated Future and Events in Memory," *American Behavioral Scientist* 29 (1986): 695–709.

49. Kathy Charmaz, *Good Days, Bad Days: The Self in Chronic Illness and Time* (New Brunswick, N.J.: Rutgers University Press, 1991), p. 4. Not all of the symbolic interactionist literature concerns temporal experience. Another major theme is a microsociological version of Durkheim's interest in time and social order. See, for example, Carl J. Couch, "Temporality and Paradigms of Thought," *Studies in Symbolic Interaction* 4 (1982): 1–33; Michael A. Katovich and Carl J. Couch, "The Nature of Social Pasts and Their Use as Foundations for Situated Action," *Symbolic Interaction* 15 (1992): 25–47; Stanford M. Lyman and Marvin B. Scott, "On the Time Track," in *A Sociology of the Absurd* (New York: Appleton-Century-Crofts, 1970); Robert P. Snow and Dennis Brissett, "Pauses: Explorations in Social Rhythm," *Symbolic Interaction* 9 (1986): 1–18; William A. Reese and Michael A. Katovich, "Untimely Acts: Extending the Interactionist Conception of Deviance," *Sociological Quarterly* 30 (1989): 159–84.

50. Carl J. Couch, "Symbolic Interaction and Generic Sociological Principles," *Symbolic Interaction* 7 (1984): 10.

51. David R. Maines, "Culture and Temporality," *Cultural Dynamics* 2 (1989): 114.

52. It is not my purpose to discuss or even assemble all of this literature, and, in any event, others have already performed these services with admirable diligence. The interested reader should consult the older, but still useful review by Paul Fraisse, "Perception

and Estimation of Time," *Annual Review of Psychology* 35 (1984): 1–36, as well as the more recent bibliography by Samuel L. Macey, *Time: A Bibliographic Guide* (New York: Garland, 1991).

53. Joseph E. McGrath and Janice R. Kelly, *Time and Human Interaction: Toward a Social Psychology of Time* (New York: Guilford, 1986), p. 68.

54. James, *The Principles of Psychology*, Vol. 1, p. 624, italics in original.

55. Mark Baker, *Cops: Their Lives in Their Own Words* (New York: Simon and Schuster, 1985), p. 117.

56. Robert E. Ornstein, *On the Experience of Time* (Baltimore: Penguin, 1969), p. 40.

57. Ibid., p. 43.

58. James B. Taylor, Louis A. Zurcher, and William H. Key, *Tornado: A Community Responds to Disaster* (Seattle, Wash.: University of Washington Press, 1970), p. 62.

59. Ornstein, *On the Experience of Time*, p. 44.

60. H. Wayne Hogan, "A Theoretical Reconciliation of Competing Views of Time Perception," *American Journal of Psychology* 91 (1978): 423.

61. Ibid., p. 423, italics in original.

62. H. Wayne Hogan, "Time Perception and Stimulus Preference as a Function of Stimulus Complexity," *Journal of Personality and Social Psychology* 31 (1975): 34.

63. Ibid., p. 33.

64. Ornstein, *On the Experience of Time*, p. 20.

NOTES TO CHAPTER 2

1. Gary King, Robert O. Keohane, and Sidney Verba, *Designing Social Inquiry: Scientific Inference in Qualitative Research* (Princeton: Princeton University Press, 1994), p. 44.

2. John Van Maanen, *Tales of the Field: On Writing Ethnography* (Chicago: University of Chicago Press, 1988), pp. 113–14.

3. James, *The Principles of Psychology*, Vol. 1, p. 624; Albert

Camus, *The Stranger*, trans. S. Gilbert (New York: Vintage, 1946), p. 100.

4. Arthur Koestler, *Dialogue with Death* (New York: Macmillan, 1960), pp. 119–20, italics in original.

5. Hogan, "A Theoretical Reconciliation of Competing Views of Time Perception."

6. Koestler, *Dialogue with Death*, p. 120.

7. Ibid., p. 119.

8. Ansel Adams, *Ansel Adams: An Autobiography* (Boston: Little, Brown, 1985), p. 7.

9. Elie Wiesel, *Night*, trans. Stella Rodway (New York: Bantam, [1958] 1982), p. 34, italics in original.

10. Isaac Asimov, *The End of Eternity* (New York: Bantam, 1955), p. 89.

11. Durkheim, *The Elementary Forms of the Religious Life*. As George Kubler observes, "Time has categorical varieties"; in other words, there are "different kinds of duration." See *The Shape of Time: Remarks on the History of Things* (New Haven: Yale University Press, 1962), pp. 83–84.

12. According to Edward Hall, "Time compresses when it speeds up." However, he muddies the water by citing an example in which an informant states, "'Everything went into slow motion.'" His terminology confuses protracted duration and temporal compression. See *The Dance of Life*, p. 125.

13. Carl Jung uses the term "synchronicity" in reference to a meaningful coincidence. It should be obvious that I define this term in a very different fashion. See Carl G. Jung, ed., *Man and His Symbols* (Garden City, N.Y.: Doubleday, 1964).

14. Anselm L. Strauss, *Mirrors and Masks: The Search for Identity* (San Francisco: Sociology Press, 1969).

NOTES TO CHAPTER 3

1. Goffman, *Frame Analysis*.

2. A fortiori, the same is true of synchronicity.

3. Emile Durkheim, *Suicide*, ed. G. Simpson, trans. J. A. Spaulding and G. Simpson (Glencoe, Ill.: Free Press, [1897] 1951).

4. Goffman, *Frame Analysis*, p. 14.

5. Zerubavel, *Hidden Rhythms*, p. 22.

6. David A. Karp and William C. Yoels, *Symbols, Selves, and Society: Understanding Interaction* (New York: J. B. Lippincott, 1979), p. 11.

7. Kristin Luker, *Abortion and the Politics of Motherhood* (Berkeley: University of California Press, 1984), p. 105, italics in original.

8. In related fashion, although with less detail, one prospective soldier had this to say of Beast Barracks, the initial weeks of training for new cadets at West Point: "'Days are so long here, these six weeks seem like a year.'" See Michael Winerip, "The Beauty of Beast Barracks," *New York Times Magazine*, 12 October 1997, p. 64.

9. Donald Knox, *Death March: The Survivors of Bataan* (New York: Harcourt Brace Jovanovich, 1983), p. 237.

10. Three television commercials provide further evidence. From 1985: the announcer asks, "Does your antacid work fast enough?" A woman responds, "It seems to take forever." From 1986: "When the sun's really high, a day of hard work seems to last a week." And, from 1994: "Every second of an asthma attack can feel like an eternity."

11. Receiving terrible news is another variation on offensive experiences: "In his mind everything moved in slow motion, the words of his father stretched out so slowly: that-he-would-die-from-AIDS because he had never heard of anyone who didn't die from it, that he would continue to work for as long as he could, that he would do whatever he could for them." See Roy Peter Clark, "Ace of Spades," *St. Petersburg Times*, Sunday, 18 February 1996, p. 3A.

12. Norman K. Denzin, *On Understanding Emotion* (San Francisco: Jossey-Bass, 1984).

13. Jack Katz, *Seductions of Crime: Moral and Sensual Attractions in Doing Evil* (New York: Basic Books, 1988), p. 31.

14. J. Segal and Z. Segal, "How to Handle Embarrassment," *Seventeen*, April 1980, p. 190.

15. Maggie Scarf, *Unfinished Business: Pressure Points in the Lives of Women* (Garden City, N.Y.: Doubleday, 1980), p. 347.

16. Tom Clancy, *The Hunt for Red October* (New York: Berkeley Books, 1985), p. 120.

17. Mary Sturt, *The Psychology of Time* (New York: Harcourt, Brace, 1925), p. 89.

18. Billy G. Gunter, "The Leisure Experience: Selected Properties," unpublished paper, Department of Sociology, University of South Florida, Tampa, 1986, p. 14.

19. Laura Flynn McCarthy, "'My First Kiss,'" *Seventeen*, February 1985, p. 138.

20. Peter L. Berger, *A Rumor of Angels: Modern Society and the Rediscovery of the Supernatural* (Garden City, N.Y.: Anchor, 1970), p. 58, italics in original.

21. Associated Press, "'All You Feel Is Wind,'" *St. Petersburg Times*, Monday, 29 August 1994, p. 3A.

22. E. B. Fein, "Earthquake Toll Staggers Soviets," *St. Petersburg Times*, Saturday, 10 December 1988, p. 8A.

23. See "River Levees Strain, Burst as Downpours Persist," *St. Petersburg Times*, Saturday, 10 July 1993, p. 6A.

24. Goffman, *Frame Analysis*, p. 562.

25. The same is true of professional race car drivers: "'You're talking about a car traveling more than the length of a football field per second. One second to go 130 yards. But in an accident, time slows down. You remember everything that happens if you don't lose consciousness from the impact. It seems like an eternity.'" See Bruce Lowitt, "Surviving the Crash," *St. Petersburg Times*, Sunday, 21 May 1995, p. 8C. Nonetheless, like those who survive natural disasters, the victims of accidental violence sometimes find the experience weirdly exhilarating. One of our informants compared her automobile accident to "a carnival ride."

26. Baker, *Cops*, p. 163.

27. David Olinger and Bob Port, "Home Gun Business Supplies Attacker," *St. Petersburg Times*, Sunday, 27 June 1993, p. 9A.

28. Jack Abbott, *In the Belly of the Beast: Letters from Prison* (New York: Random House, 1981), p. 76.

29. C. Sutton, "USF Police Officer's Date Rape Seminar Is Eye-Opener for Men," *St. Petersburg Times*, Tuesday, 28 June 1988, p. 8.

30. Thomas French, "A Drive through Overtown," *St. Petersburg Times*, Saturday, 17 March 1984, p. 1A. Robbery can have a similar effect on the perceived passage of time:

> "I said, 'Take whatever you want, but don't hurt anybody,'" the 28-year-old waitress recalled Tuesday. "The gun was on my neck the whole time. It felt like an hour. It felt like it would never end."

See Kelly Ryan, "Worker Wounded in IHOP Robbery," *St. Petersburg Times*, Wednesday, 18 December 1996, p. 1B.

31. Associated Press, "Tourists Jump Gunman, Turn into Heroes." *St. Petersburg Times*, Monday, 31 October 1994, p. 10A. The source of the danger need not be interpersonal violence. Consider the following description of one passenger's escape from the wreckage of an airplane:

> From under a pile of people, luggage and seats torn from their runners, Barrington, the personnel manager, glimpsed daylight. He struggled to reach, then unsnap his seat belt.
>
> "It probably took 15 seconds. For that eternity, I was sure that I would be trapped and burned," he said.

See Dan Sewell, "In Crash, Minutes Are a Lifetime," *St. Petersburg Times*, Friday, 25 August 1995, p. 9A.

32. Michael G. Flaherty, "Reality Play: A Sociological Analysis of Amusement" (Ph.D. diss., University of Illinois at Urbana-Champaign, 1982).

33. William Hauptman, "In Tornado Alley," *St. Petersburg Times*, Sunday, 22 April 1984, p. 23A.

34. James, *The Principles of Psychology*, Vol. 1, p. 626.

35. Lance Morrow, "Waiting as a Way of Life," *Time*, 23 July 1984, p. 65.

36. Patricia Bosworth, *Diane Arbus: A Biography* (New York: Avon, 1985), p. 80.

37. William Shakespeare knew that, when properly handled, delay can add a piquant, if protracted tension to consummation:

Come now; what masques, what dances shall we have,

To wear away this long age of three hours

Between our after-supper and bed time?

See "A Midsummer Night's Dream," in *The Annotated Shakespeare*, Vol. 1, *The Comedies*, ed. A. L. Rowse (New York: Clarkson N. Potter, [1594] 1978), p. 271.

38. Christopher Smart, "Pair Drifts in Gulf 20 Miles after Catamaran Flips," *St. Petersburg Times*, Wednesday, 3 October 1984, p. 3B.

39. Leo Tolstoy, *Anna Karenina*, Vol. I., trans. C. Garnett (New York: Random House, [1877] 1939), p. 844.

40. Barry Schwartz, *Queuing and Waiting* (Chicago: University of Chicago Press, 1975), p. 168.

41. Chaikin, *A Man on the Moon*, pp. 108–109. A reporter attributes a similar experience to a soldier who cannot get his weapon to fire during an important demonstration:

For 90 seconds that must have seemed an eternity, the anonymous camouflaged infantryman checked out his weapon while a dozen top Soviet and American generals stared at him in silence. Finally he got the portable, wire-guided missile to fire— and scored a bull's-eye on a wooden target nearly 2000 feet away.

See Associated Press, "Powell Handles Soviet Gaffe with Customary Cool," *St. Petersburg Times*, Wednesday, 24 July 1991, p. 11A.

42. Clancy, *The Hunt for Red October*, p. 75.

43. Marilyn Beck, "8 Weeks Here like 8 Years to 'Cocoon' Cast Member," *St. Petersburg Times*, Wednesday, 10 October 1984, p. 1D.

44. Jack Anderson, "The Twins of Auschwitz Today," *Parade Magazine*, Sunday, 2 September 1984, p. 6.

45. Abbott, *In the Belly of the Beast*, p. 44.

46. Malcolm Braly, *False Starts: A Memoir of San Quentin and*

Other Prisons (Boston: Little, Brown, 1976), p. 181, italics in original.

47. Tim Roche, "In Jail, You Bide Your Time by Filling It," *St. Petersburg Times*, Sunday, 28 March 1993, p. 8B.

48. Jim Carrier, "Eagle-Tracking Carries Researchers to Heights, Depths," *St. Petersburg Times*, Wednesday, 11 December 1985, p. 24A.

49. Edward T. Hall, *The Silent Language* (New York: Doubleday, 1959), p. 179.

50. Joseph Heller, *Catch-22* (New York: Simon and Schuster, 1961), p. 9.

51. Mihaly Csikszentmihalyi, *Beyond Boredom and Anxiety* (San Francisco: Jossey-Bass, 1975).

52. Thomas De Quincey, *Confessions of an English Opium-Eater*, ed. M. Elwin (London: MacDonald, [1821] 1956), p. 314.

53. Aldous Huxley, *The Doors of Perception* (New York: Harper and Row, 1954), p. 21.

54. Victor Gioscia, *TimeForms* (New York: Interface, 1974), p. 7. According to C. R. Carroll, the "major characteristics of the psychedelic state" include a "heightened awareness of sensory input, experienced as a flood of sensation," as well as "diminished control over what is experienced." See *Drugs in Modern Society* (Dubuque, Iowa: William C. Brown, 1989), p. 272. The interested reader should also consult Stephens Newell, "Chemical Modifiers of Time Perception," as well as Frances E. Cheek and Joan Laucius, "The Time Worlds of Three Drug-Using Groups—Alcoholics, Heroin Addicts, and Psychedelics," in *The Future of Time*, ed. H. Yaker, H. Osmond, and F. Cheek (Garden City, N.Y.: Doubleday, 1971).

55. Murray S. Davis, *Smut: Erotic Reality/Obscene Ideology* (Chicago: University of Chicago Press, 1983), p. 20.

56. Ibid., p. 72.

57. Hal Lipper, "'Empire' a Masterful Epic," *St. Petersburg Times*, Friday, 11 December 1987, p. 6.

58. Sturt, *The Psychology of Time*, p. 110.

59. James S. Gordon, "The UFO Experience," *Atlantic*, August 1991, p. 88.

60. Betty Malz, *My Glimpse of Eternity* (Waco, Tex.: Chosen Books, 1977), p. 88.

61. See *U.S. News and World Report*, "The Near-Death Experience—How Thousands Describe It," 11 June 1984, p. 59. In his book, Ring quotes one of his respondents saying, "it seems like it was forever." See Kenneth Ring, *Life at Death: A Scientific Investigation of the Near-Death Experience* (New York: Coward, McCann, and Geoghegan, 1980), p. 97. Likewise, another student of near-death experiences quotes an informant as saying, "It seemed then as though time were standing still." See Raymond A. Moody, Jr., *Life after Life: The Investigation of a Phenomenon—Survival of Bodily Death* (Marietta, Ga.: Mockingbird, 1975), p. 46.

62. Tom Robbins, *Jitterbug Perfume* (New York: Bantam, 1984), p. 322. The supernatural is frequently signaled by the transcendence of temporality. Note, for instance, the use of multiple tenses when Catholics recite the "mystery of faith" during their mass: "Christ has died, Christ is risen, Christ will come again." See James Socias, ed., *Daily Roman Missal* (Princeton, N.J.: Scepter, 1994), p. 727.

63. Eldridge Cleaver, *Soul on Ice* (New York: McGraw-Hill, 1968), p. 11.

64. Morag Coate, "The Cosmic Crises," in *In Their Own Behalf: Voices from the Margin*, ed. C. H. McCaghy, J. K. Skipper, Jr., and M. Lefton (New York: Appleton-Century-Crofts, 1968), p. 66.

65. Erving Goffman, *Encounters: Two Studies in the Sociology of Interaction* (Indianapolis, Ind.: Bobbs-Merrill, 1961), p. 38.

66. Michael Murphy and John Brodie, "'I Experience a Kind of Clarity,'" *Intellectual Digest* 3 (1973): 19–20. Apparently, receivers feel the same way:

> "When you come open like that, everything slows down," Mills said. "It's like time stops. You stand there thinking, 'Come on, come on, get me the ball.' But it got there."

See John Romano, "Steelers Outlast Miami 24–17," *St. Petersburg Times*, Tuesday, 26 November 1996, p. 1C.

67. Ernest K. Gann, "How Kathy Rude Won Her Race with Death," *Parade Magazine*, Sunday, 9 December 1984, p. 6.

68. Bruce Lowitt, "Reggie! Reggie!" *St. Petersburg Times*, Thursday, 2 June 1994, p. 1C.

69. Frank Deford, "The Energy Bunny," *Newsweek*, 22 December 1997, p. 80.

70. B. J. Phillips, "Finishing First, At Last," *Time*, 13 August 1984, p. 46.

71. Joan Didion, *The White Album* (New York: Simon and Schuster, 1979), p. 83.

72. Associated Press, "Rescue Unit Travels 400 Miles into Atlantic to Save Crew of Indian Freighter," *St. Petersburg Times*, Sunday, 19 February 1984, p. 6B.

73. Fyodor Dostoevsky, *Crime and Punishment*, trans. C. Garnett (New York: Random House, [1866] 1956), p. 69.

74. Virginia Woolf, *Orlando* (New York: Harcourt, Brace, 1928), p. 98.

75. Hall, *The Dance of Life*, p. 19.

76. Charles Darwin, *The Life and Letters of Charles Darwin*, Vol. 1, ed. F. Darwin (New York: Basic Books, [1888] 1959), p. 29.

77. Arlie Russell Hochschild, *The Managed Heart: Commercialization of Human Feeling* (Berkeley: University of California Press, 1983), p. 30.

78. Malcolm Jones, "Private Lives," *St. Petersburg Times*, Wednesday, 22 August 1984, p. 1D.

79. Mark Baker, *Nam: The Vietnam War in the Words of the Men and Women Who Fought There* (New York: William Morrow, 1981), p. 164.

80. Jennifer L. Stevenson, "'Blue' Results Baffle Franz," *St. Petersburg Times*, Tuesday, 13 September 1994, p. 8C.

81. Jan Glidewell, "Businessman Finds Joy in This Four-Wheel Toy." *St. Petersburg Times*, Sunday, 2 October 1988, p. 8.

82. Felix Hess, "The Aerodynamics of Boomerangs," *Scientific American* 219 (1968): 127.

83. In the following vignette, the narrator refers to "horror," but the rest of her story leaves one with the impression that shock is really at issue:

On one morning dive, I found two green- and yellow-striped nudibranches of a kind I had not see [*sic*] before. My husband, Jim, was trying out his new underwater video camera, but the habits of a still photographer were strong in him, so he posed the two pretty shell-less snails on a shelf of rock to photograph them together. Suddenly—to our horror—the larger of the two shot out a huge cerulean-blue sleeve—a mouth half as big as itself—and tried to envelope the other. Just in time, the smaller one thrust out its own sleeve like a spinnaker sail, making itself too large to be engorged. For minutes the two of them wrestled, each one trying to wrap the other in the ballooning folds of its own gorgeous mouth. It seemed an eternity before the two creatures fell away from each other.

See Frances Fitzgerald, "The Galápagos, Caravansary of the Sea," *New York Times Magazine, The Sophisticated Traveler*, Sunday, 15 September 1996, pp. 23–24.

84. Beirne Keefer, "Of a Marlin and Magic," *St. Petersburg Times*, Tuesday, 13 November 1984, p. 1C.

85. Hall, *The Dance of Life*, p. 130.

86. David R. Maines, "On Time, Timing, and the Life Course," *Interdisciplinary Topics in Gerontology* 17 (1983): 187.

87. Thomas Mann, *The Magic Mountain*, trans. H. T. Lowe-Porter (New York: Alfred A. Knopf, 1968), pp. 104–105.

88. In 2 percent of the cases, there was not enough information to classify the episode.

NOTES TO CHAPTER 4

1. Schutz and Luckmann, *The Structures of the Life-World*, p. 56.

2. J. David Lewis and Andrew J. Weigert, "The Structures and Meanings of Social Time," *Social Forces* 60 (1981): 437.

3. Hogan, "A Theoretical Reconciliation of Competing Views of Time Perception," p. 417.

4. As a measure of reliability, we coded 94.7 percent of the narratives into the same categories.

5. Mead, *Mind, Self, and Society*, p. 83.

6. Ibid., p. 94.

7. Ibid., p. 122.

8. Mead, *The Philosophy of the Act*, p. 79.

9. John Dewey, *Human Nature and Conduct* (New York: Henry Holt, 1922), p. 178. Elsewhere, Dewey states that emotion is provoked when "the conditions for mere habit are denied." See "The Theory of Emotion, II: The Significance of Emotions," *Psychological Review* 2 (1895): 27.

10. Mead, *Mind, Self, and Society*, p. 149.

11. James, *The Principles of Psychology*, Vol. 1, p. 239.

12. James Boswell, *The Life of Samuel Johnson* (New York: Modern Library, [1791] 1931), p. 725.

13. Ralph H. Turner and Victoria Billings, "The Social Contexts of Self-Feeling," in *The Self-Society Dynamic: Cognition, Emotion, and Action*, ed. J. A. Howard and P. L. Callero (Cambridge: Cambridge University Press, 1991), p. 119, italics in original.

14. Goffman, *Frame Analysis*, p. 339.

15. Schutz and Luckmann, *The Structures of the Life-World*, p. 3.

16. Jonathan H. Turner, *A Theory of Social Interaction* (Stanford: Stanford University Press, 1988), p. 49.

17. Schutz and Luckmann, *The Structures of the Life-World*, p. 35.

18. Goffman, *Frame Analysis*, p. 345.

19. Mead, *Mind, Self, and Society*, p. 93n.

20. Ulric Neisser, *Cognition and Reality: Principles and Implications of Cognitive Psychology* (San Francisco: W. H. Freeman, 1976).

21. Schutz and Luckmann, *The Structures of the Life-World*, p. 60.

22. Schwartz, *Queuing and Waiting*, p. 168. The individual's involvement with his or her circumstances is, in this instance, a response to an unusual absence of overt activity.

23. A salesclerk told us about a night at "work" so bereft of activity that it seemed as if there was only the beating of her own heart to mark the passage of time.

24. Goffman, *Encounters*, pp. 38–39.

25. Stanford W. Gregory, Jr., "A Quantitative Analysis of Temporal Symmetry in Microsocial Relations," *American Sociological Review* 48 (1983): 129, italics in original.

26. Herbert Blumer, *Symbolic Interactionism: Perspective and Method* (Englewood Cliffs, N.J.: Prentice-Hall, 1969), p. 17.

27. Simmel, "The Metropolis and Mental Life," p. 328.

28. Lewis and Weigert, "The Structures and Meanings of Social Time," p. 439.

29. Harold Garfinkel, *Studies in Ethnomethodology* (Englewood Cliffs, N.J.: Prentice-Hall, 1967), p. 53.

30. Goffman, *Frame Analysis*, p. 8.

31. Schutz and Luckmann, *The Structures of the Life-World*, p. 3.

32. Ibid., p. 18. Their use of this concept is derived from the earlier work of American pragmatists. See John Dewey, *Problems of Men* (New York: Philosophical Library, 1946); William James, *Pragmatism* (New York: Longmans, Green, 1907); and Charles Sanders Peirce, *Collected Papers of Charles Sanders Peirce*, Vol. 5, ed. C. Hartshorne and P. Weiss (Cambridge: Harvard University Press, 1960).

33. This is in keeping with Goffman's observation that "routines of social intercourse in established settings allow us to deal with anticipated others without special attention or thought." See *Stigma: Notes on the Management of Spoiled Identity* (Englewood Cliffs, N.J.: Prentice-Hall, 1963), p. 2.

34. Aaron V. Cicourel, *Cognitive Sociology: Language and Meaning in Social Interaction* (New York: Free Press, 1974), p. 11.

35. Goffman, *Encounters*, p. 80.

36. Ibid., pp. 39–40. Goffman reiterates his argument in *Frame Analysis*: "[I]t is often understood that although a particular degree of involvement is preferred, considerable variation in intensity is

acceptable, boredom marking one boundary and 'overinvolvement' the other." See p. 117.

37. Goffman, *Frame Analysis*, p. 345.

38. Hogan, "A Theoretical Reconciliation of Competing Views of Time Perception."

39. Hand-to-hand combat and solitary confinement were cited earlier as examples, but the U-shaped curve should be conceived of as continuous. Therefore, less divergent incidents, such as minor medical emergencies and periods of self-conscious waiting, have the same kind of effect, albeit to a lesser degree.

40. Temporal compression is not synonymous with that which Csikszentmihalyi has termed "flow." His concept concerns the subjective and situated conditions for enjoyment, and he finds that "people who enjoy what they are doing enter a state of 'flow': they concentrate their attention on a limited stimulus field, forget personal problems, lose their sense of time and of themselves, feel competent and in control, and have a sense of harmony and union with their surroundings." See *Beyond Boredom and Anxiety*, p. 182. The experience of temporal compression occurs not only in enjoyable circumstances but also in unpleasant circumstances, as evidenced by the passage from Koestler's autobiography in chapter 2. See *Dialogue with Death*, pp. 119–20. Indeed, a majority of the people in my narratives are not enjoying themselves.

41. Koestler, *Dialogue with Death*, p. 120.

42. Mark H. Ashcraft, *Human Memory and Cognition* (Glenview, Ill.: Scott, Foresman, 1989), pp. 67–68.

43. Ibid., p. 68.

44. Ibid., p. 68.

45. And, a fortiori, less challenging tasks require even less conscious attention, at least up to a certain point, past which boredom makes the task problematic and redirects attentional resources to self and situation, thereby producing the experience of protracted duration.

46. Ashcraft, *Human Memory and Cognition*, p. 241.

47. Koestler, *Dialogue with Death*.

NOTES TO CHAPTER 5

1. Walter Wallace, *The Logic of Science in Sociology* (Chicago: Aldine-Atherton, 1971).

2. Earl Babbie, *The Practice of Social Research*, 5th ed. (Belmont, Calif.: Wadsworth, 1989), p. 45.

3. Norman K. Denzin, *The Research Act: A Theoretical Introduction to Sociological Methods*, 3d ed. (Englewood Cliffs, N.J.: Prentice-Hall, 1989), p. 50.

4. King, Keohane, and Verba, *Designing Social Inquiry*, p. 190.

5. Babbie, *The Practice of Social Research*; Denzin, *The Research Act*; Eugene J. Webb, Donald T. Campbell, Richard D. Schwartz, Lee Sechrest, and Janet Belew Grove, *Nonreactive Measures in the Social Sciences*, 2d ed. (Boston: Houghton Mifflin, 1981).

6. King, Keohane, and Verba, *Designing Social Inquiry*, p. 3.

7. Bear in mind that 705 narratives served as the basis for the analysis in chapter 3. Altogether, then, the experiences of 1,071 people are represented in the qualitative and quantitative data.

8. Thus, it is not a simple random sample from a known sampling frame. However, Ramon Henkel points out that "our data can be thought of as a random sample from some hypothetical universe composed of data . . . like those at hand." See *Tests of Significance* (Beverly Hills, Calif.: Sage, 1976), p. 85. As is typically the case with procedures of this sort, I assume that my statistical tests address the following question: "[I]f such a population were to exist, could sampling error account for any differences between the expected and observed value of the statistic?" Ibid., p. 86.

9. The word "less" refers to the total volume (i.e., number) of experiences per standard temporal unit.

10. Charles E. Collyer and James T. Enns, *Analysis of Variance: The Basic Designs* (Chicago: Nelson-Hall, 1987).

11. SPSS, *SPSS-X User's Guide*, 3d ed. (Chicago: SPSS, 1988).

12. As David Maines reminds us, there are "patterns of life events in relation to age." See "On Time, Timing, and the Life Course," p.

186. It follows that the density of experience per standard temporal unit is also structured by "age norms."

13. Carol D. Ryff, "The Subjective Experience of Life-Span Transitions," in *Gender and the Life Course*, ed. A. S. Rossi (New York: Aldine, 1985), p. 105.

14. John P. Robinson, "The Time Squeeze," *American Demographics* 12 (1990): 33.

15. Bertram J. Cohler, "Personal Narrative and Life Course," in *Life-Span Development and Behavior*, Vol. 4, ed. P. B. Baltes and O. G. Brim, Jr. (New York: Academic Press, 1982), p. 221.

16. David L. Gutmann, "Parenthood: A Key to the Comparative Study of the Life Cycle," in *Life-Span Developmental Psychology: Normative Life Crises*, ed. N. Datan and L. H. Ginsberg (New York: Academic Press, 1975), p. 170; See also idem, "The Cross-Cultural Perspective: Notes Toward a Comparative Psychology of Aging," in *Handbook of the Psychology of Aging*, ed. J. E. Birren and K. W. Schaie (New York: Van Nostrand Reinhold, 1977), p. 312; and idem, "The Post-Parental Years: Clinical Problems and Developmental Possibilities," in *Mid-Life: Developmental and Clinical Issues*, ed. W. H. Norman and T. J. Scaramella (New York: Brunner/Mazel, 1980), p. 44.

17. Carnegie Corporation of New York, *A Matter of Time* (New York: Carnegie Corporation of New York, 1992), p. 28.

18. Irving Rosow, *Socialization to Old Age* (Berkeley: University of California Press, 1974), p. 9.

19. Gutmann, "The Post-Parental Years," p. 40.

20. Fine, "Organizational Time."

21. Csikszentmihalyi, *Beyond Boredom and Anxiety*.

22. Csikszentmihalyi's book is based on quantitative as well as qualitative data. However, the chapter on surgery reports only qualitative data. See *Beyond Boredom and Anxiety*.

23. Fine, "Organizational Time," p. 107.

24. Ibid., p. 106.

25. Ibid., p. 109.

26. Here, I refer to my discussion in chapter 4 concerning the relationship between "routine complexity" and "automatic processing."

27. Fine, "Organizational Time," p. 109.

28. One of our informants provides corroborating evidence:

On Fridays, we are very busy at the bank, and the lobby stays open until 6:00. Usually, it closes at 4:00. I had gotten to work early that morning and was constantly helping customers without a break except for a half lunch. At about 3:30, the office was a *chaos* because the workers from Ark Constructions all came in to cash their checks. Everyone was super busy trying to get to people without taking too much time. It seemed as if the rush would never end; it was just one customer after another. Your fingers don't stop flying and you don't stop moving because you don't want anyone mad at you. It seems like we didn't get everyone out of there . . . until a half hour later. In actuality, it was only five minutes. [italics added]

Her use of the word "chaos" suggests that she finds a particular segment of the rush problematic. Thus, her experience of protracted duration is in keeping with what my theory would predict.

29. Fine, "Organizational Time," p. 107. The phrase "too many external demands" does *not* refer to the rush per se, but to the problematic rush—that is, one in which the chefs are either overwhelmed by the sheer quantity of demands or one in which things go wrong. The phrase "not enough" refers to the opposite of the rush—that is, an interval during which there are not enough demands, and the chefs are bored.

30. Csikszentmihalyi, *Beyond Boredom and Anxiety.*

31. Ibid., p. 127.

32. Ibid., p. 130.

33. Ibid., p. 130.

34. Ibid., p. 132.

35. Ibid., p. 132.

36. Ibid., pp. 43 and 86.

37. Given widespread use of variations on the word "automatic," it may well be that the concept "automatic processing" (chapter 4) has almost as much emic currency as etic.

38. Blumer, *Symbolic Interactionism*, pp. 21–22.

NOTES TO CHAPTER 6

1. This is, of course, a line from "The Walrus and the Carpenter," by Lewis Carroll. See *The Complete Works of Lewis Carroll* (New York: Vintage, [1872] 1976), p. 186.

2. Randall Collins, "Toward a Neo-Meadian Sociology of Mind," *Symbolic Interaction* 12 (1989): 2.

3. Ibid., p. 18.

4. Denzin, *On Understanding Emotion*, p. 58.

5. Katz, *Seductions of Crime*, p. 31.

6. Charmaz, *Good Days, Bad Days*, p. 90.

7. Fine, "Organizational Time"; Csikszentmihalyi, *Beyond Boredom and Anxiety*.

8. Goffman, *Encounters*, p. 39.

9. Clifford Geertz, *The Interpretation of Cultures* (New York: Basic Books, 1973), p. 432.

10. Mead, *Mind, Self, and Society*, p. 37.

11. Goffman, *Frame Analysis*.

12. Thomas J. Scheff, "Toward Integration in the Social Psychology of Emotions," *Annual Review of Sociology* 9 (1983): 333–54; Peggy A. Thoits, "The Sociology of Emotions," *Annual Review of Sociology* 15 (1989): 317–42; Peter L. Callero, "Toward a Sociology of Cognition," in *The Self-Society Dynamic: Cognition, Emotion, and Action*, ed. Judith A. Howard and Peter Callero (New York: Cambridge University Press, 1991); Cicourel, *Cognitive Sociology*; Zerubavel, *The Fine Line*; idem, *Social Mindscapes*.

13. Lindesmith, Strauss, and Denzin, *Social Psychology*, p. 94.

14. James, *The Principles of Psychology*, Vol. 1, p. 239, italics in original.

15. For the most part, I restrict this review to studies that make time a central topic of discussion. There are, however, any number of other anthropological studies that mention temporality in passing, as when A. R. Radcliffe-Brown notes the integration of botany and time-reckoning among the Andaman Islanders:

> In the jungles of the Andamans it is possible to recognize a distinct succession of odours during a considerable part of the year as one after another the commoner trees and lianas come into flower. . . . The Andamanese have therefore adopted an original method of marking the different periods of the year by means of the different odoriferous flowers that are in bloom at different times. Their calendar is a calendar of scents.

See *The Andaman Islanders* (New York: Free Press, [1922] 1964, pp. 311–12.

16. Martin P. Nilsson, *Primitive Time-Reckoning* (Lund: C. W. K. Gleerup, 1920). There are representatives from other disciplines who concur with this anthropological emphasis on cultural relativity. In psychology, Robert Levine has shown that "the pace of life" varies across cultures. See "The Pace of Life," *American Scientist* 78 (1990): 450; see also Robert V. Levine and Ellen Wolff, "Social Time: The Heartbeat of Culture," *Psychology Today* 19 (March 1985): 28–35. Three sociologists—Mark Iutcovich, Charles E. Babbitt, and Joyce Iutcovich—have arrived at much the same conclusion: "Within any society and/or culture, people may demonstrate different temporal orientations and therefore perceive punctuality in a variety of ways." See "Time Perception: A Case Study of a Developing Nation," *Sociological Focus* 12 (1979): 71–85.

17. Irving A. Hallowell, "Temporal Orientation in Western Civilization and in a Pre-Literate Society," *American Anthropologist* 39 (1937): 647.

18. E. E. Evans-Pritchard, "Nuer Time-Reckoning," *Africa* 12 (1939): 189–216; Paul Bohannan, "Concepts of Time among the Tiv of Nigeria," *Southwestern Journal of Anthropology* 9 (1953): 251–62.

19. Benjamin Lee Whorf, *Language, Thought, and Reality* (Cambridge, Mass.: M.I.T. Press, 1956), p. 58.

20. Ibid., p. 55.

21. Murray Wax, "The Notions of Nature, Man, and Time of a Hunting People," *Southern Folklore Quarterly* 26 (1962): 182; T. O. Beidelman, "Kaguru Time Reckoning: An Aspect of the Cosmology of an East African People," *Southwestern Journal of Anthropology* 19 (1963): 9.

22. Alfonso Ortiz, *The Tewa World: Space, Time, Being, and Becoming in a Pueblo Society* (Chicago: University of Chicago Press, 1969); Douglas R. Givens, *An Analysis of Navajo Temporality* (Lanham, Md.: University Press of America, 1977); Trudy Griffin-Pierce, *Earth Is My Mother, Sky Is My Father: Space, Time, and Astronomy in Navajo Sandpainting* (Albuquerque: University of New Mexico Press, 1992); Gary H. Gossen, *Chamulas in the World of the Sun: Time and Space in a Maya Oral Tradition* (Prospect Heights, Ill.: Waveland, 1984).

23. Susan U. Philips, "Warm Springs 'Indian Time': How the Regulation of Participation Affects the Progression of Events," in *Explorations in the Ethnography of Speaking*, 2d ed., ed. R. Bauman and J. Sherzer (Cambridge: Cambridge University Press, 1989), p. 92.

24. Susan Sontag, *Against Interpretation* (New York: Farrar, Straus, and Giroux, 1966), p. 14.

25. Florence Rockwood Kluckhohn, "Dominant and Variant Value Orientations," in *Personality in Nature, Society, and Culture*, 2d ed., ed. Clyde Kluckhohn, Henry A. Murray, and David M. Schneider (New York: Alfred A. Knopf, 1959), p. 348.

26. Hall, *The Dance of Life*, p. 4.

27. Ibid., p. 126.

28. Hoyt Alverson, *Semantics and Experience: Universal Metaphors of Time in English, Mandarin, Hindi, and Sesotho* (Baltimore: Johns Hopkins University Press, 1994), p. ix.

29. Ibid., pp. ix and xi.

30. Ibid., p. xii.

31. Much the same problem arises in the study of culture and personality. See Ruth Benedict, *Patterns of Culture* (Boston: Houghton Mifflin, 1934).

32. Durkheim, *The Elementary Forms of the Religious Life*, p. 492.

33. Clyde Kluckhohn and Henry A. Murray, "Personality Formation: The Determinants," in *Personality in Nature, Society, and Culture*, 2d ed., ed. Clyde Kluckhohn, Henry A. Murray, and David M. Schneider (New York: Alfred A. Knopf, 1959), p. 53.

34. Peter L. Berger and Thomas Luckmann, *The Social Construction of Reality* (Garden City, N.Y.: Anchor, 1967).

35. Erving Goffman, *Interaction Ritual: Essays on Face-to-Face Behavior* (Garden City, N.Y.: Anchor Doubleday, 1967), p. 44.

36. Norman K. Denzin, "Symbolic Interactionism and Ethnomethodology: A Proposed Synthesis," *American Sociological Review* 34 (1969): 926.

37. Maines, "Culture and Temporality," p. 117.

38. Italo Calvino, *t zero*, trans. William Weaver (San Diego: Harcourt Brace Jovanovich, 1969), p. 95.

39. Lillian Schlissel, *Women's Diaries of the Westward Journey* (New York: Schocken, 1982), p. 228.

40. Ping Yao, "Power, Resistance, and Accommodation: Affect in a Women's Writing System in Jiangyong, China," unpublished manuscript, Department of Anthropology, University of Illinois at Urbana-Champaign, 1993, p. 26.

41. Arlie Russell Hochschild, "Emotion Work, Feeling Rules, and Social Structure," *American Journal of Sociology* 85 (1979): 551–75; idem, *The Managed Heart*.

42. Hochschild, *The Managed Heart*, p. 43.

43. Hochschild, "Emotion Work, Feeling Rules, and Social Structure," p. 552.

44. James R. Averill and Elma P. Nunley, *Voyages of the Heart: Living an Emotionally Creative Life* (New York: Free Press, 1992), p. 9.

45. Howard S. Becker, "History, Culture, and Subjective Experi-

ence: An Exploration of the Social Bases of Drug-Induced Experiences," *Journal of Health and Social Behavior* 8 (1967): 164.

46. Clinton R. Sanders, *Customizing the Body: The Art and Culture of Tattooing* (Philadelphia: Temple University Press, 1989).

47. James, *The Principles of Psychology*, Vol. 1, p. 402, italics in original. I am indebted to Scott Harris for bringing this statement to my attention. See "Status Inequality and Close Relationships: An Integrative Typology of Bond-Saving Strategies." *Symbolic Interaction* 20 (1997): 4.

48. Mead, *Mind, Self, and Society*, p. 25. As a psychiatric variation on the same theme, Harry Stack Sullivan conceives of "selective inattention" as the "power of the self-system to control focal awareness." See *The Interpersonal Theory of Psychiatry* (New York: W. W. Norton, 1953), p. 233.

49. Mead, *Mind, Self, and Society*, pp. 103 and 99.

50. Ibid., p. 98.

51. George Gonos, "'Situation' Versus 'Frame': The 'Interactionist' and the 'Structuralist' Analyses of Everyday Life," *American Sociological Review* 42 (1977): 854–67. However, Goffman's empirical observations are not in accord with the more deterministic principles that are espoused in his programmatic statements. See Michael G. Flaherty, "Two Conceptions of the Social Situation: Some Implications of Humor," *Sociological Quarterly* 31 (1990): 93–106.

52. Erving Goffman, *The Presentation of Self in Everyday Life* (Garden City, N.Y.: Anchor Doubleday, 1959), p. 9.

53. Goffman, *Encounters*, p. 104.

54. Abraham H. Maslow, *Toward a Psychology of Being*, 2d ed. (New York: D. Van Nostrand, 1968), p. 3.

55. Ibid., p. 13.

56. Ibid., pp. 32 and 92.

57. Csikszentmihalyi, *Beyond Boredom and Anxiety*, pp. 3 and 5.

58. Katz, *Seductions of Crime*, p. 216.

59. Ibid., p. 3.

60. Elizabeth G. Menaghan, "Social Psychology: Interpreting the Data," *Social Psychology Quarterly* 58 (1995): 323.

61. Ben Hamper, *Rivethead: Tales from the Assembly Line* (New York: Warner, 1992), p. 95.

62. Ibid., p. 97.

NOTES TO METHODOLOGICAL APPENDIX

1. Michael G. Flaherty, "The Neglected Dimension of Temporality in Social Psychology," *Studies in Symbolic Interaction* 8 (1987): 143–55; idem, "Multiple Realities and the Experience of Duration," *Sociological Quarterly* 28 (1987): 313–26; idem, "The Perception of Time and Situated Engrossment," *Social Psychology Quarterly* 54 (1991): 76–85; idem, "The Erotics and Hermeneutics of Temporality," in *Investigating Subjectivity: Research on Lived Experience*, ed. Carolyn Ellis and Michael G. Flaherty (Newbury Park, Calif.: Sage, 1992); "Conceptualizing Variation in the Experience of Time," *Sociological Inquiry* 63 (1993): 394–405; idem, "Some Methodological Principles from Research in Practice: Validity, Truth, and Method in the Study of Lived Time," *Qualitative Inquiry* 2 (1996): 285–99; Michael G. Flaherty and Michelle D. Meer, "How Time Flies: Age, Memory, and Temporal Compression," *Sociological Quarterly* 35 (1994): 705–21.

2. John M. Johnson and David L. Altheide, "Reflexive Accountability," *Studies in Symbolic Interaction* 11 (1990): 25.

3. John M. Johnson and David L. Altheide, "The Ethnographic Ethic," *Studies in Symbolic Interaction* 14 (1993): 101.

4. Yvonna S. Lincoln and Norman K. Denzin, "The Fifth Moment," in *Handbook of Qualitative Research*, ed. Yvonna S. Lincoln and Norman K. Denzin (Thousand Oaks, Calif.: Sage, 1994), p. 578.

5. George E. Marcus, "What Comes (Just) after 'Post'? The Case of Ethnography," in *Handbook of Qualitative Research*, ed. Yvonna S. Lincoln and Norman K. Denzin (Thousand Oaks, Calif.: Sage, 1994), pp. 568–69.

6. Denzin, *The Research Act*; John Lofland and Lyn H. Lofland, *Analyzing Social Settings* (Belmont, Calif.: Wadsworth, 1995); Anselm L. Strauss, *Qualitative Analysis for Social Scientists* (Cam-

bridge: Cambridge University Press, 1987); Robert M. Emerson, ed., *Contemporary Field Research: A Collection of Readings* (Prospect Heights, Ill.: Waveland, 1988); Paul C. Higgins and John M. Johnson, eds., *Personal Sociology* (New York: Praeger, 1988); M. Patricia Golden, ed., *The Research Experience* (Itasca, Ill.: F. E. Peacock, 1976).

7. But see Carl J. Couch, "Why I Went into the Laboratory: And, What We Found," *Studies in Symbolic Interaction* 16 (1994): 21–34; Norman K. Denzin, "Researching Alcoholics and Alcoholism in American Society," *Studies in Symbolic Interaction* 11 (1990): 81–101; Janet Finch and Jennifer Mason, "Decision Taking in the Fieldwork Process: Theoretical Sampling and Collaborative Working," in *Studies in Qualitative Methodology* 2 (1990): 25–50.

8. Murphy and Brodie, "'I Experience a Kind of Clarity,'" pp. 19–20.

9. Howard S. Becker, *Outsiders: Studies in the Sociology of Deviance* (New York: Free Press, [1963] 1973), p. 189.

10. Ibid., p. 42.

11. Each time that evidence turned up where I expected it was tantamount to provisional confirmation of my working hypothesis. These discoveries bolstered my confidence in the emerging theory.

12. Of course, there is a sense in which all data are "generated" by someone, if only by the curious sociologist reading the popular press. What is more, we must bear in mind that "found" data have been generated by others—reporters, editors, autobiographers, and the like—and usually with a very different set of relevancies.

13. Webb et al., *Nonreactive Measures in the Social Sciences*, p. 118.

14. Huxley, *The Doors of Perception*; Camus, *The Stranger*; Didion, *The White Album*.

15. Barney G. Glaser and Anselm L. Strauss, *The Discovery of Grounded Theory* (Chicago: Aldine, 1967), p. 61.

16. Denzin, *The Research Act*, p. 28; Webb et al., *Nonreactive Measures in the Social Sciences*, p. 35.

17. Norman K. Denzin, "The Logic of Naturalistic Inquiry," *Social Forces* 50 (1971): 166–82.

18. Over time, my coding of the data had produced six broad themes: suffering and intense emotions, violence and danger, waiting and boredom, altered states, concentration and meditation, shock and novelty. Eventually, it became evident that new interviews no longer required reformulation of these categories.

19. Webb et al., *Nonreactive Measures in the Social Sciences*, pp. 34–35.

20. Sanders, *Customizing the Body*, p. viii.

21. James, *The Principles of Psychology*, Vol. 1, p. 624, italics in original.

22. Goffman, *Frame Analysis*, p. 1.

23. Florian Znaniecki, *The Method of Sociology* (New York: Farrar and Rinehart, 1934).

24. Alfred R. Lindesmith, "Comment on W. S. Robinson's 'The Logical Structure of Analytic Induction,'" *American Sociological Review* 17 (1952): 492.

25. Becker, *Outsiders*, p. 45, italics in original.

26. Peter K. Manning, "Analytic Induction," in *Qualitative Methods*, ed. R. B. Smith and P. K. Manning (Cambridge, Mass.: Ballinger, 1982), p. 277, italics in original.

27. Alfred R. Lindesmith, *Addiction and Opiates* (Chicago: Aldine, [1947] 1968), p. 10.

28. Denzin, "Symbolic Interactionism and Ethnomethodology: A Proposed Synthesis," p. 926. This goal is in keeping with Mead's epistemology. In *Mind, Self, and Society* (p. 34), Mead argues that the researcher "wishes to make these statements as universal as possible, and is scientific in that respect."

29. Glaser and Strauss, *The Discovery of Grounded Theory*, p. 101.

30. Blumer, *Symbolic Interactionism*, p. 22.

31. Turner and Billings, "The Social Contexts of Self-Feeling," p. 119.

Csikszentmihalyi, *Beyond Boredom and Anxiety.*

33. Interdisciplinary research is one aspect of a pragmatic orientation.

34. Becker, *Outsiders,* p. 203. Likewise, Ralph Turner suggests that we stand to gain from integrating analytic induction with quantitative methods. See "The Quest for Universals in Sociological Research," *American Sociological Review* 18 (1953): 604–11.

35. Jeffrey E. Nash makes a similar point with his "opportunistic sociology." See "It's Good for 'Em: Object Lessons in Youth Sports," in *Personal Sociology,* ed. Paul C. Higgins and John M. Johnson (New York: Praeger, 1988), p. 14.

36. Everett C. Hughes, *The Sociological Eye* (Chicago: Aldine Atherton, 1971), p. 453. I am indebted to Anselm Strauss for bringing this essay to my attention. See "Everett Hughes: Sociology's Mission," *Symbolic Interaction* 19 (1996): 271–83.

37. Marcus, "What Comes (Just) after 'Post'?" p. 571.

38. Becker, *Outsiders,* p. 198.

39. John Van Maanen, *Tales of the Field: On Writing Ethnography* (Chicago: University of Chicago Press, 1988).

40. Mead, *Mind, Self, and Society.* So it is that one conducts research with a particular audience in mind, just as one writes with a particular audience in mind.

41. Ludwik Fleck, *Genesis and Development of a Scientific Fact,* trans. F. Bradley and T. J. Trenn (Chicago: University of Chicago Press, [1935] 1979), p. 25.

42. Opposition takes the form of those who are "both anti-science and radically non-empirical." See Sheldon Stryker, "Progress? Well, Yes and No: The States of Sociological Social Psychology," *Social Psychology Quarterly* 58 (1995): 329.

43. Webb et al., *Nonreactive Measures in the Social Sciences,* p. 34.

44. Fleck, *Genesis and Development of a Scientific Fact,* p. 2.

45. Harvey A. Farberman, "The Foundations of Symbolic Interaction: James, Cooley, and Mead," *Studies in Symbolic Interaction,* Supplement 1 (1985): 13–27. As Alfred Lindesmith points out,

progress is inherent in the use of analytic induction: "The progressive refinement of theory that is brought about by the necessity of taking negative instances seriously regardless of their frequency also makes for a progressively closer articulation of theory with the empirical evidence generated by research." See *Addiction and Opiates*, p. 20. Given the inevitable limits to the individual's time and effort, even the most assiduous study leaves us with little more than working hypotheses. However, inquiry is a collective enterprise, and, over time, the work of many researchers moves us forward.

46. Couch also refers to "generic sociological principles," a phrase that echoes an earlier statement by Erving Goffman: "Since alienation can occur in regard to any imaginable talk, we may be able to learn from it something about the generic properties of spoken interaction." See Couch, "Symbolic Interaction and Generic Sociological Principles," pp. 1 and 10; Goffman, *Interaction Ritual*, p. 114.

Index

Abbott, Jack, 53, 60

Activity, 9, 25; alternative lines of, 6; cessation of ordinary, 56; checked, 89; embedded, 29, 94, 97, 99; Mead's subordination of experience to, 7; overt, 21, 24, 28–29, 32, 34, 38, 91, 94, 97, 112, 130, 198n. 22; problematic, 138; repetitive, 138; social structure of, 101; tacit, 90, 94; usual forms of, 14; willful, 74; without conscious involvement, 119

Adams, Ansel, 30

Age: as independent variable, 121–22, 125–28; and patterns of life events, 201n. 12; and perceived passage of time, 79, 110; of subjects, 119–20

Agency, 62, 152–59. *See also* Choice; Intentionality; Self-determination; Volition; Will

Altered states: daydreams as, 65; deprivation and, 67; dreams as, 64–65, 68; drugs and, 63–64, 68, 152–53, 194n. 54; exertion and, 67; fantasy as, 64–65, 68; nervous breakdown as, 68; protracted duration and, 63–69; psychotic interlude as, 68–69; rapture as, 67, 72; sexuality as, 49, 64, 68, 77; stress and, 67; stupor as, 67–68; as a sufficient cause, 43, 81–82 table 1; UFO sightings as, 66, 68

Altheide, David L., 161

Alverson, Hoyt, 146

Analysis: descriptive, 135; inductive, 171; of process and meaning, 169–70; and working hypotheses, 85, 170, 210n. 11, 213n. 45. *See also* Logic; Methods

Anthropology of time, 143–44, 146, 205nn. 15, 16

Anticipation, 9, 59

Arousal, 93

Ashcraft, Mark H., 106

About the Author

Michael G. Flaherty is Professor and Chair of Sociology at Eckerd College. He received his Ph.D. from the University of Illinois at Urbana-Champaign. He is the coeditor (with Carolyn Ellis) of *Investigating Subjectivity: Research on Lived Experience* and *Social Perspectives on Emotion*, Volume 3. Currently, he is the editor of *Symbolic Interaction*, and he serves as an advisory editor for *Social Psychology Quarterly*, *Sociological Quarterly*, and *Time and Society*.